Fuels
and the
National Policy

John N. McDougall

Butterworths
Toronto

Fuels and the National Policy
© 1982—Butterworth & Co. (Canada) Ltd.

Printed and bound in Canada
5 4 3 2 1   2 3 4 5 6 7 8 9/8

Canadian Cataloguing in Publication Data
McDougall, John N.
    Fuels and the national policy

Bibliography: p.
Includes index.
ISBN 0-409-84805-0

1. Energy policy – Canada – History.   2. Energy
policy – Canada.   3. Energy development – Canada –
History.   4. Fossil fuels – Canada.   I. Title.

HD9502.C32M32   333.8'215'0971   C82-094412-2

The Butterworth Group of Companies

Canada:
Butterworth & Co. (Canada) Ltd., Toronto and Vancouver

United Kingdom:
Butterworth & Co. (Publishers) Ltd., London

Australia:
Butterworths Pty. Ltd., Sydney

New Zealand:
Butterworths of New Zealand Ltd., Wellington

South Africa:
Butterworth & Co. (South Africa) Ltd., Durban

United States:
Butterworth (Publishers) Inc., Boston
Butterworth (Legal Publishers) Inc., Seattle
Mason Publishing Company, St. Paul

Contents

For Doris McDougall

Preface

I AM INDEBTED to four sources of inspiration for the over-all concept and fundamental orientation of this volume. One of these, going back to the very inception of the project, was an incidental observation made by John Smart, who wondered what I would make of the possibility that something resembling the "Ottawa Valley Line" might have existed for the Canadian coal market in the 1920s. This got me thinking about the possibility that a number of parallels could be found between national policies with respect to coal, when it was the primary source of energy for the country, and those concerning oil and natural gas, when they became the chief sources. My approach to this and other past issues in Canada's energy policy was influenced further by Hugh Thorburn's article concerning the role of Parliament in the Pipeline Debate of 1956, where he made the point that while Parliament does not govern—or, as we would say today, "make policy"—it still constitutes "our most outstanding body of experts on local and regional public opinion," whose role it is to give voice to the desires and interests of Canadians across the country. This observation reinforced my interest in undertaking a comprehensive reading of the Commons record on Canada's fuel problems, and inspired my hope that the major Commons debates on international and interprovincial trade in fuels would at least reveal the conflicting interests and considerations bearing on the formulation of Canadian energy policies, even if they might not by themselves provide an explanation of why these policies came into being. A third influence was the writing of George Grant, whose lament for Canada took on particular meaning for me as I watched the Northern Pipeline Act, and hence the Alaska Highway Pipeline, win the support in principle of all three of Canada's national political parties, as if combining with the United States in the largest private venture in history were nothing but a matter of factory orders in central Canada, and devoid of substantial (and for me, decisive) ramifications for Canadian independence in anything more than the most emasculated, "legal" sense. Finally, there was Hannah Arendt's succinct yet momentous challenge to us all to "think what we are doing," as we stand poised (to act, perhaps?) between past and future.

I have accumulated an intimidating number of more tangible debts in the process of bringing this work to publication. First and foremost in the realm of practical assistance was a research grant from the Canada Council, without which I could not have managed the archival work I was able to do in Edmonton, Ottawa, and Halifax, or have drawn on the research assistance of Coulson Askew. I am very grateful for this support, and I hope that after all this time, the SSHRC is able to regard this disbursement on the part of its predecessor as money well spent. I am grateful to Coulson Askew for permitting me to include in chapters 4 and 7, material he originally presented in his master's thesis on the party politics of pipeline development in Canada. François Bregha, Andrew Sancton, and Robert Young went considerably out of their way to provide me with comments and suggestions that proved invaluable in giving final shape to the manuscript. Successive versions of the manuscript were processed by the Social Science Computing Lab under the immediate supervision of Joanne Stewart, who did more than I had a right to expect in helping me to proceed through successive drafts at a respectable pace. Isabel Shearer and Janet Horowitz patiently and conscientiously provided auxiliary typing services. Finally, Martin Westmacott, Chairman of the Department of Political Science and Robert Hohner, Associate Dean of Social Science, have my thanks for taking the trouble to arrange special funding for incidental expenses.

Between the conception of this project and its final realization falls the usual shadow—in this case myself, whose failings and limitations are solely responsible for the fact that, despite all of the above-mentioned help, errors, misjudgements, and problems remain.

Permission was gratefully received to incorporate into chapters 5, 6 and 7 material that was originally published in "Regulation versus Politics: The National Energy Board and the Mackenzie Valley Pipe Line," Andrew Axline, et al., eds., *Continental Community? Independence and Integration in North America* (Toronto: McClelland and Stewart, 1974); and "Prebuild Phase or Latest Phase? The United States Fuel Market and Canadian Energy Policy" (1981) *International Journal* 36, 1.

John N. McDougall
The University of Western Ontario
July 1982

1   The Quest for a National Fuel Policy

DURING the 1970s, many of the most controversial issues in Canadian politics had to do with various aspects of Canadian energy policy: the Mackenzie Valley and Alaska Highway pipelines, Petro-Canada, natural gas exports, and the "two-price" system, to mention just a few. It would take pages to list, let alone discuss, the various considerations that were drawn into the debates that took place over these matters, but there was one aspect common to them all. The goal of national self-sufficiency in fuels played some part in all of these major energy debates, and it played a central part in many of them. In this respect, the recent debates on energy policy in Canada resembled major energy debates of the past. The idea that the government of Canada should take measures whereby the entire Canadian fuel market would be served exclusively from Canadian fuel sources has been with us since Confederation; and calls for a "national fuel policy" along these lines have been heard, with varying intensity, from different regions of the country throughout our history. This study is an attempt to explain why these calls for a national fuel policy to promote energy self-sufficiency for Canada have persisted over the years; why their intensity has grown or diminished at different times in different regions; and why successive federal governments have been willing or unwilling to adopt policies encouraging the Canadian consumption of Canadian fuels.

Self-sufficiency as a Goal of National Energy Policies

Since the Arab oil embargo of 1973, every major oil-importing country has given a high priority to the goal of reducing its dependence on crude oil imports from the Middle East and, in the longer run, of achieving national energy self-sufficiency.[1] The measures undertaken by different countries in the direction of these objectives have varied, largely according to the type and availability of indigenous energy sources. For the countries of Western Europe and North America, the main responses have been to substitute local fuel sources (such as natural gas and coal) for imported oil in the short term, and to promote the development of new alternative sources of energy in the longer term. Thus, Canada shares with the United States and Western Europe a general strategy of, first, reducing as far as possible its dependence on Middle Eastern oil by securing alternative sources of imported oil; second, reducing its need for imported oil in general by promoting the consumption of alternative fuels, especially natural gas; and third, reducing its demand for fossil fuels in general by developing other means of generating electrical power and providing heat. To this extent, Canada's commitment to the goal of national self-sufficiency in energy had its origin in the same world developments as other countries' commitments to the same goal, and the general direction of its approach to the realization of self-sufficiency is comparable to that of other industrialized countries. In one respect, however, Canada's present concern to reduce as far as possible its dependence on imported fuel and to adopt national policies to promote the consumption of domestic fuels is different.

For most industrialized countries, national vulnerability to disruptions in the supply of foreign fuel sources is relatively recent, a direct consequence of a number of significant developments in their patterns of energy consumption since World War II. One of these developments was the availability of cheap Middle Eastern oil between the end of World War II and the Arab oil embargo of 1973. Prior to World War II, the industrial growth of these countries had been fueled by coal, and the market for petroleum products—while it had been expanding rapidly throughout the century—was still confined largely to transportation uses. Following World War II, the comparatively cheaper price and greater convenience of petroleum products gradually brought about the substitution of oil for coal as the primary fuel source in most national energy markets. In Europe and Japan, this conversion from coal to oil translated directly into a rapidly increasing demand for imported oil, owing to a lack of indigenous sources of petroleum.[2] In North America, this conversion resulted initially in the expansion of both oil imports and the continental oil and gas industry. Along with the rise in the consumption of fuel oils came a rapid increase in the demand for transportation fuels. Countries that had been traditionally self-sufficient in fuels found themselves relying more and more heavily on cheap foreign oil, while at the same time erecting barriers to the importation of foreign oil in order to protect their national coal industries (in the case of Western Europe) or their domestic coal and petroleum industries (in the case of North America). Despite these measures, all the industrialized countries of the West had lost their "energy self-reliance" by 1973, to the extent that a suspension or interruption in the flow of oil from the Middle East could threaten disaster (or at least severe inconvenience) for a large proportion of their population through industrial shutdowns and a scarcity of fuels for space heating and transportation.

Canada shared in all of these postwar developments, and has also responded to recent disruptions in its foreign fuel supply with policies aimed to increase its use of domestically produced fuels. The difference between Canada and other countries is that Canada has always been subject to disruptions in its fuel supply or, to put it another way, has never been self-sufficient in fuels. Consequently, policies relating to the import, export and (especially) transportation of fuels have figured prominently in Canadian energy policy from Confederation to the present. In fact, Canadian energy policy can be summed up in a single sentence: national energy policies are (and always have been) centrally concerned with fuels, and Canadian fuel policy is (and always has been) centrally concerned with the transportation of fuels. The reasons for the prominence of fuel transportation in Canadian energy policy are to be found, first, in the difference between the two basic ways in which energy is used in a society—energy consumed as power and energy consumed as heat—and second, in the geographic distribution of Canada's primary energy sources in relation to its major centres of energy consumption.

Heat Versus Power
In providing for all of their material needs, human beings are constantly, and necessarily, either putting and keeping things in motion or heating them up. In pre-industrial times, we obtained the power to move things by means of the exertion of human or animal labour and through the harnessing of natural forces, such as wind and falling water. For example, a millstone might be turned by a man on a treadmill, an ox on a treadmill, a windmill, or a water wheel. In those times, we obtained our capacity to make things melt, boil, and react chemically with one another through the burning of fuel—dung, wood, coal, oils. In the more recent industrial age, we have found our source of motive power in the steam engine, electric motor, and internal combustion engine, while we have continued to rely on fuels—mainly coal, petroleum and manufactured or natural gas—as sources of heat. Correspondingly, it is useful to divide a market for energy (especially the market for energy in an advanced industrial country) into its two major components: the demand for motive power to drive the machinery of industry, and the demand for heat to produce the transformation processes essential to the manufacture of the goods consumed in an advanced society. (Other uses of energy, such as illumination, are not ignored in this analysis, but are excluded because they represent only a small portion of total energy consumption.) Until recently, both the power and the heat requirements of industrial societies were met through the use of coal, and the availability of coal became a necessary condition of large-scale industrialization.[3]

During the industrial expansion that took place in the latter half of the nineteenth century, the power required for manufacturing and transportation was translated into a demand for fuels (primarily coal) by the medium of the steam engine. In addition, coal was required for a variety of direct-heat processes such as the production of iron and steel, and was used increasingly for space-heating purposes in homes, factories, and offices (usually replacing wood). As a consequence, access to reliable supplies of coal was an indispensable requirement of large-scale industrialization. The high cost of transporting coal (especially over land) meant that proximity to large deposits of coal was a highly significant factor in the location of industry and, secondarily, of urbanization. Accordingly, most industrialized countries were self-sufficient in coal; if they had not had reliable access to adequate supplies of coal, they would not have become industrialized in the first place. Finally, with the advent of electrical power and the replacement of the steam engine by the electric motor, the industrial demand for power continued to translate into demand for coal for the thermal generation of electricity, except in those countries and regions blessed with the capacity for hydroelectric production. The availability of coal therefore remained a significant factor in the location and expansion of manufacturing industries, although the capacity for hydroelectric power production became a factor in the location of light industry and some metallurgical industries, where the cost of electrical power represented a significant proportion of the total cost of production.

The Geography of Energy in Canada
The foregoing observations concerning the distinction between heat and power and their respective parts in the energy consumption of an industrial society are an important starting point in the understanding of some of the distinctive features of Canada's energy picture, and hence some of the fundamental factors shaping Canadian energy policy. Historically, Canadian energy policy has been preoccupied with issues relating to the trade and transportation of fuels because of a fundamental difference in the geography of fuels and hydroelectric power in Canada. Central Canada (defined roughly as the St. Lawrence Lowlands and the Great Lakes basin) is the location of Canada's highest demand for energy in all its forms. During the twentieth century, this region was able to develop large local hydroelectric power resources, but it has always been deficient in local fuels, and has had to rely instead on fuels imported from foreign sources or from distant parts of Canada. In fact, it is probably fair to say that the early industrialization of what is now Ontario was contingent upon the region's access to nearby deposits of American coal. At the same time, Quebec met its coal needs through water-borne imports from foreign sources or from the Maritime provinces of Nova Scotia and New Brunswick. Subsequently, the fuel needs of the region (including, in part, the use of fuels for local thermal generation of electricity) continued to be met through imports of coal, oil, and natural gas from elsewhere, while its power needs have been supplied locally through hydroelectric and thermal-electric production within the two central provinces themselves. As a result, interprovincial

and international trade in electrical power in Canada has been histori- cally less troubled than interprovincial and international trade in fuel sources, which has been a matter of absolute necessity, constant uncer- tainty, and periodic crisis. Moreover, since international and inter- provincial trade come under federal jurisdiction in Canada, the energy problems of central Canada have required action on the part of the federal government with respect to fuels far more frequently than with respect to electrical power; and this action inevitably affected the in- terests of other provinces as owners of fuel resources, and the interests of other countries (particularly the United States) as both sources of sup- ply for Canadian consumers and potential markets for Canadian pro- ducers of fuel.

Toward a National Fuel Policy
Having become industrialized initially on the basis of imported coal, central Canada has always been heavily dependent on foreign fuels, and concerns about the reliability of supply have frequently given rise to the idea of substituting Canadian energy sources for imported fuels; that is, the idea of making energy self-sufficiency the primary objective of na- tional energy policy. While the development of hydroelectric power— "white coal"—unquestionably played its part in the movement to deliver Canada from its "bondage to American coal" during the early decades of this century, the idea of promoting the substitution of Cana- dian fuel sources for imported fuels has been a persistent theme from the Confederation debates of 1865 to the Northern Pipeline debate of the late 1970s.[4] Not even an unlimited supply of hydroelectric power would have reduced the demand for fuels in central Canada to zero; and to the extent that the availability of power did encourage the growth of certain industries there, it also contributed indirectly to an increase in the general energy requirements of the growing economy, many of which could only be met through the use of coal, oil, or gas. Hence, the desirability and the feasibility of adopting a national fuel policy, whereby the entire Canadian market for fuels would be served ex- clusively from Canadian sources of supply stand as the most enduring and most controversial questions relating to energy policies at the na- tional level in Canada.

It would be misleading to suggest that, throughout their history, Canadians have debated the pros and cons of energy self-sufficiency as such. In fact, widespread usage of the term seems to be of relatively re- cent vintage. What Canadians *have* debated historically is the desirabil- ity (or otherwise) of adopting a "national" fuel policy—or, indeed, a "national fuel policy"—which has clearly been understood by both its supporters and its opponents to mean a set of federal measures designed to expand the use of Canadian fuels in Canadian markets, or (what amounts to the same thing) to reduce the use of imported fuels in Canadian markets. Whereas any policy adopted by the federal government, even one of completely unrestricted international trade in fuels, could properly be labelled a "national" policy merely by virtue of being a policy of the national government, it is clear that the "national policy" concept in Canada's energy debates has carried larger connotations. One such connotation harks back to the notions of nation building and national unity that were bound up in John A. Macdonald's original National Policy: interprovincial trade in fuels, like interprovincial trade in all goods, contributes to the unity and general prosperity of the country. If Canada is to be "one country," it must create and maintain a national market—national in the sense that it is "nation-wide," and national in the sense that it is not "continental" or North American. It is along these lines, for example, that the construction of the Trans-Canada natural gas pipeline has been compared with the construction of the Canadian Pacific Railway. A second connotation involves related but distinct implications of the supposed "national interest" in avoiding dependence on foreigners for something as vital as the country's fuel supply: a national energy policy is not one adopted merely in order to bring particular benefits to this or that group of Canadians, but rather to shore up the independence of the country as a whole and to eliminate the possibility that other countries might wring concessions on matters of national concern by threatening to cut off the fuel supply of a substantial portion of the country. A national policy must promote "all-Canadian" solutions to Canadian problems.

Given that in these basic ways the proponents of a national fuel policy have been able to infuse the creation of a nationwide fuel market with notions of such symbolic and tangible importance to the country as its very unity and independence, it seems particularly significant that they have never seen the policy come about. The most obvious generalization that one can make about the history of Canadian energy policy is that federal governments have consistently gone part of the way, but never all of the way, toward the creation of a national fuel market. That is, they have repeatedly adopted measures to expand the Canadian consumption of Canadian fuels beyond what it would otherwise have been, but never to the point of the complete substitution of Canadian fuels for foreign fuels. Both the extremes of unrestricted imports of fuels and the complete exclusion of imported fuels have been avoided by federal governments throughout our history, which suggests constant tensions and compromises between political forces and national considerations that would tend to push national policies in either direction.

The record of Canada's energy debate clearly reveals a set of such conflicting interests and considerations, as well as a series of trade-offs in national policy that reflects them: conflict between the political benefits and economic costs of a national fuel market; between the consumer interest in low-cost supplies of fuel and in reliable supplies of fuel; between the producer interest in more lucrative foreign markets and in more secure national ones; between the consumer interest in both reliable and low-cost supplies and the producer interest in both secure and lucrative markets; and finally, perhaps, between those two overriding goals—national unity and national independence. The history of energy policy in Canada shows that Canadian diversity is so great that even the highest of national purposes must benefit most at the expense of some, or benefit some at the expense of most, with the fate of the nation constantly suspended between a transcendent national interest and conflicting regional interests. As if this were not enough, the federal government has also had to respond to different American priorities with respect to the export and import of coal, oil and natural gas.

One obvious reason, then, for the persistence of calls for a national fuel policy is that the entire Canadian market has never been served by domestic sources of supply for any fuel. In other words, the national policy concept has never been fully implemented, and Canadian governments have regularly come under criticism from those who were impressed by the fact—as they saw it—that Canada had the capacity to meet its own fuel needs, but was not doing so. Because Canada has almost always exported a large share of the fuels it has produced while importing a large share of the fuels it has consumed, there has been some force to the notion that Canada should act to reduce, if not eliminate, fuel imports, and to increase in turn Canada's consumption of its own fuel supplies—in other words, to reduce international trade in fuels and to expand interprovincial trade.

For over one hundred years, Canadian governments have faced policy decisions regarding the extent to which they ought to promote interprovincial as opposed to international trade in fuels—from the first coal tariff of 1870 to the Trans-Quebec and Maritimes Pipeline of 1981. Taken together, the measures they have adopted or rejected to give effect to such decisions have always been central elements of Canadian energy policy. Similarly, the vast majority of Canada's energy debates over the same period have been carried on between coalitions of interests attempting either to promote or to discourage more active government intervention in the direction of national self-sufficiency. It may not be quite accurate to describe Canada's energy debate as an ongoing dialogue between nationalists promoting the "national policy" concept of maximum interprovincial trade and continentalists promoting a free-market concept of unhindered international trade (although some of the individual debates along the way have come very close to this), but it is fair to say that nearly every major debate this country has had concerning the trade and transportation of Canadian fuel sources has been centrally concerned with policies which would take the country nearer to one or the other of these two poles. Of course, merely to observe that government responses to these conflicting pressures over the years have consistently fallen somewhere between these two possible policy extremes is not to explain why this is so, and the remainder of this study is devoted to a review of national fuel policies since Confederation and the various factors that account for them. The rest of the present chapter will be devoted to an outline of the major developments in Canadian fuel policy, to be examined in detail in later chapters.

Unity, Independence and Self-sufficiency, 1867–1980

Perhaps the simplest explanation for the failure of successive Canadian governments to fully adopt a policy of national self-sufficiency in fuels, despite persistent pressure upon them to do so, is that to do so has always seemed too costly. As C. D. Howe once explained his refusal to endorse such a policy with respect to coal: "The government cannot change the geography of Canada, nor can it ignore the immutable laws of economics."[5] Howe here brings to mind an analysis of Canadian economic history in which Canadians are told that their economy has been shaped by three determinants—the physical environment, the laws of economics, and government policy—and further that government policies have had lasting effect on the economy only to the extent that they have not cut across the grain of those physical and economic constraints.[6] Howe would not have had trouble making the case that the physical and economic constraints bearing on the delivery of Canadian coal to central Canadian markets were prohibitive; and what was then referred to as "Canada's Fuel Problem" consisted precisely in the fact that central Canada, which then (as now) contained the preponderant share of the country's population and industry, had no indigenous coal supplies, while the bulk of Canada's coal was located in the distant provinces of Nova Scotia, Alberta, and British Columbia.[7] The cost of overcoming these constraints was so high compared with that of importing coal from the United States that it does indeed seem more reasonable to ask why a national coal policy should even have been attempted than to ask why such a policy was never completely realized. Yet moves in the direction of such a policy were made repeatedly with respect to coal, and have continued with respect to oil and gas. This leads to the question, What circumstances and considerations above those of comparative cost have prompted Canadian governments to intervene in one way or another to expand Canada's consumption of its own sources of fuel instead of relying on lower-priced imported supplies?

One answer to this question is that national self-sufficiency on the one hand and national unity on the other have been valued substantially, though not absolutely, throughout the history of the country. This answer would be consistent with the rhetoric commonly employed by Canadians promoting a "national fuel policy" aimed at achieving national self-sufficiency in fuels. The advocates of self-sufficiency have identified that objective so closely with the independence and unity of the nation that its only partial realization historically would appear to represent a failure on the part of the government to act in the national interest. However, since those who have advocated the adoption of a national policy have tended to do so only when it would benefit their own region or sector at the expense of others, it is more appropriate to view government policy as a compromise between such conflicts of interest, or as a partial concession to particular regional and sectoral demands, than as a lack of commitment to the national good. Nevertheless, the ends of national unity and independence have historically been (and continue to be) such a central part of Canada's political debate concerning the creation of a nationwide market for the country's fuels that it is important to understand how the proponents of a national fuel policy have argued that it would serve these ends, as well as to appreciate the circumstances in which different interests have put such arguments forward.

With respect, then, to the objective of independence in Canadian energy policy, it would appear that as in so many areas of Canada's national life, the country's proximity to the United States has given a peculiar twist to the relationship between energy self-sufficiency and national independence. Any country is justified in paying the extra cost of eliminating its dependence on foreign sources of fuels to the extent that the nation values either the security of its fuel supply or the independence of its foreign policy; and, as we shall see, Canadians have been reminded several times of the dangers of leaving a substantial segment of the country's population and economy to rely on other countries for their fuel needs. But Canadians have also experienced the conflict between the political benefits and economic costs of self-sufficiency in another, more fundamental form, since the economically rational solution to the Canadian fuel problem—the unrestricted importation of foreign fuels and the unrestricted exportation of Canadian fuels to foreign markets—has time and again proven impossible politically because of the independent existence of the country as such. As one student of the Canadian coal situation in the 1930s put it, the *sine qua non* of the Canadian fuel problem was Canada's status as an independent country, for were Canada and the United States, for example, one country, both politically and economically, there would be no move to push Alberta coals into Ontario . . . There would be no move to push Maritime Province coals into Eastern Ontario at all.[8]

The Canadian problem reflected in this assessment is that the most accessible supply of coal, oil, and natural gas for Ontario and Quebec has always been the United States, whereas those regions of the United States within a short distance of Canada's fuel producers have always been the natural market for Canadian fuels.[9] Consequently, national self-sufficiency in fuels has never seemed as attractive economically as unrestricted international trade, and federal measures to promote the interprovincial movement of Canadian fuels have generally been adopted when the United States has been either unwilling or unable to make American fuels fully available to Canadian markets or to allow Canadian fuel producers unrestricted access to American markets. That is, policies approaching the national policy model have been implemented to the degree that they have in order either to compensate Canadian producers for their lack of access to American markets or to protect central Canadian consumers against interruptions in foreign supplies, neither of which situations would have arisen if North America were a single economic and political unit. In this sense, the higher costs associated with a national as opposed to a continental fuel market can be seen as a premium (or a penalty) to be paid for the entire set of policies and principles that have both reinforced and reflected Canada's status as a distinct country. When this independent status has not represented an impediment to free international trade in fuels, neither producers nor consumers in Canada have lent much support to the creation of a national fuel market. Despite their arguments that energy self-sufficiency is vital to the preservation of national independence, calls for a national fuel policy have generally come from those whose interests have been harmed by the failure to achieve or maintain a satisfactory pattern of North American trade.

With respect to the objective of national unity in Canadian fuel policy, it is interesting to note that one of the earliest speeches ever made about fuels in the Canadian House of Commons argued the benefits that interprovincial trade in coal would bring to the new dominion by helping to bind together its separate regions:

If you want to make this great Confederation of British North American Provinces a Union in fact as well as a Union on paper, it is the duty of the Government—whatever political party may occupy the Treasury Benches—to consider with the greatest care and attention every means by which interprovincial trade may be fostered and promoted. Is there any great injustice in asking the people of Ontario to submit to a tax of 50 cents or 75 cents per ton on coal?[10]

This is far from the only time a similar argument has been made and a comparable question put. Indeed, Tupper's question raises in one form an issue that has been at the centre of nearly all of the great energy debates that have gripped this country: the justice of the terms of Confederation. Fuel-producing provinces pay extra in order to maintain markets for industries in central Canada that enjoy the protection of tariffs against otherwise cheaper foreign manufacturers, and they in turn demand larger markets in central Canada for their fuels. Along with the controversy over how large such compensation should be, there has been the even more contentious issue of how its cost should be met: higher prices for the central Canadian consumer? Subsidies on transportation paid out of general tax revenues? Reduced revenues to transportation companies? In sum, while an "all-Canadian" solution to the Canadian fuel problem has always been argued in principle to serve the cause of national unity, any moves toward its practical implementation have always been politically divisive because of the highly uneven regional incidence of the costs and benefits associated with it. Either producers have asked consumers to pay more than import prices for the sake of national unity and fairness, or consumers have asked producers to accept less than import prices for the same reasons. Canadians have shown a tendency to endorse the expansion of interprovincial trade in fuels only to the extent that other Canadians could be found to pay for it—and those other Canadians have generally objected to the added burden.

For example, until coal lost its place as Canada's major fuel source, roughly half the central Canadian market was reserved for Nova Scotia coal. Tariffs against United States coal were combined with transportation subventions for Nova Scotia coal in order to confine the purchase of United States coal to the central Canadian markets west of Cornwall, Ontario. Similarly, from 1960 to 1973, roughly half the central Canadian market for oil was supplied by Alberta oil, while federal regulation of the Canadian oil industry confined overseas oil to those Canadian markets east of Cornwall. In other words, with respect to both coal and oil, the federal government adopted policies that prevented imported coal and oil from encroaching upon one of the two major fuel markets in central Canada. The fact that the precise geographic designation of this limit was the Ottawa Valley in both cases may be nothing more than an interesting coincidence, but the effect of the policy in both cases was to preserve one of the two major fuel markets in the country for Canadian production. These policies toward coal and oil reflected a willingness on the part of the federal government to impose higher fuel costs on central Canadian consumers in order to protect Canadian producers who could not otherwise compete with imports into central Canada and who faced difficulty in obtaining or maintaining adequate markets in the United States.

In contrast, the fuel that has been marketed in Canada in a manner most closely approximating the national policy model is natural gas. As will be related below, the federal government instituted policies, including direct financial assistance, to ensure that both Toronto and Montreal would have access to Alberta natural gas. The most obvious explanation for this was less likely a fondness for the national policy concept in the heart of the arch-continentalist C. D. Howe than the fact that, in this instance, Howe had been informed by the United States government that continental trading arrangements were not on: American authorities had given notice that no assured long-term supplies of American natural gas would be available for Canada. As we shall see with regard to coal during the 1920s and oil after 1973, when central Canada has been unable to look outside the country for reliable supplies of fuel, it has looked to Alberta. These policies reflect the determination of the federal government to ensure that central Canadian consumers would have access to reliable supplies of fuel—a priority that, with respect to natural gas in particular, has run contrary to the expressed preference of producers in the west for higher rates of exports to the United States.

It would appear from this brief overview that despite the prominence of their role in Canadian debates on energy policy, the goals of national unity and independence have not in themselves played a large part in shaping the policies actually adopted toward the trade and transportation of Canadian fuels. More fundamental forces determining such policies and the political conflicts surrounding them would appear to be such factors as the geographic distribution of Canadian fuel supplies in relation to major centres of energy consumption and the availability of American fuel supplies and markets to Canadian consumers and producers. However, a systematic analysis of the relative significance of the determinants of Canadian fuel policies that have affected the extent to which the Canadian market has been served by domestic fuels must await the detailed examination of national fuel policy and politics since Confederation that is provided in the remaining chapters of this study. At this stage, it is sufficient to say that the implacable realities of geography and transportation economics have repeatedly surfaced as formidable obstacles to the creation of a national fuel market in Canada and the elimination of Canada's need to import fuels. In fact, this study is essentially a story of the controversies and policies of the past 114 years that have centred on the problem of overcoming these obstacles and distributing the costs and benefits associated with attempts to do so.

Of course, it is nothing new to give transportation problems a central place in an account of historical developments in Canada.[11] A preliminary *prima facie* case for doing so can be made by considering patterns in the Canadian consumption of Canadian energy sources in relation to the comparative cost of transporting different forms of energy. A 1957 study of energy in Canada reported the comparative cost of energy transportation as follows (in cents per 100 miles per one short ton of coal equivalent):[12]

Electrical power via high-tension line	316.5 to 395.5
Bituminous coal by rail	70.0 to 80.0
Bituminous coal by water	25.0 to 30.0
Petroleum via 30-inch diameter pipeline	9.0 to 13.5
Natural gas via 34-inch diameter pipeline	28.0 to 40.8

Even a crude use of these figures helps to explain a number of observations made in this study about patterns of interprovincial trade in Canadian energy sources, such as the fact that historically the interprovincial movement in electrical power has been negligible relative to such trade in fuels; the fact that Alberta oil and natural gas have been marketed in both Ontario and Quebec, while Alberta coal until recently has been almost completely shut out of even southern Ontario; the fact that Nova Scotia coal scarcely penetrated the Toronto and southwestern Ontario market even with the assistance of tariff protection and freight subventions; and the fact that Canadian fuel producers have consistently preferred closer United States markets to distant Canadian ones. This is not to say that policies of the Canadian government have been completely without effect on the international and interprovincial movement of energy sources, but it is to suggest that the geographic constraints bearing on the Canadian consumption of domestic fuel supplies and the provision of adequate Canadian markets for Canadian fuel producers have been substantial.

There is another reason, finally, for attempting to underscore the importance of transportation in the task of guaranteeing domestic supplies of fuel to all Canadians and eliminating their dependence on imported supplies: the importance of fuel transportation systems is not as widely or as fully appreciated as it could be, given the reasons for Canada's historical failure to achieve this objective and the government's recent re-commitment to the goal of energy self-sufficiency by 1990. Most Canadians appear to consider the problem of self-sufficiency as solely (though perhaps not simply) a problem of bringing Canadian productive capacity in energy into balance with energy demand (a condition the country has generally enjoyed throughout its history); they appear to be unaware of the more difficult problem of assuring that all Canadians have reliable *access* to energy produced in Canada (a condition the country has never known). In this they are not significantly helped by academic and government studies of energy in Canada that pay little or no attention to the difficulty of moving Canadian fuels from source to use. Even the federal government's recent blueprint for energy self-sufficiency, *The National Energy Program, 1980,* has much to say about the role of independence, security, fairness, accelerated exploration and development, conservation, and Canadianization in its new initiatives in the energy field; but it has almost nothing to say about the problems involved in bringing new sources of supply to major Canadian markets. This seems a curious omission, given that one of the central features of the new national program is a set of provisions to divert the exploration and development activity of the oil and gas industry away from producing areas to which Canadian consumers have long had access and toward even more distant frontier areas to which they as yet have no access. It is even more curious given that the last thirty years of oil and gas pipeline development in Canada have established quite clearly what strategy is most likely to be adopted for the enormous task of providing central Canadian markets with access to these distant domestic fuel supplies: to rely heavily on American capital, technology, and markets in the development of a continental solution to Canada's fuel problem.

2 Coal and National Unity, 1867–1913

THE STORY of federal coal policies in Canada during the years when coal was the country's principal source of energy is largely a record of various forms of assistance—freight subventions, tariffs, and miscellaneous subsidies—adopted from time to time and to varying degrees either to expand the share of the Canadian market available to Canadian producers or to protect Canadian consumers against interruptions in imports from the United States, or both. Coal did not account for more than half of Canadian energy consumption until the 1890s; thereafter, it held this position until shortly after World War II. At its peak in the early 1920s, coal provided 70 to 75 percent of the energy consumed in Canada.[1] Throughout the period, however, Canadian production rarely equalled more than half the amount of coal consumed in the country, despite complaints of shut-in capacity from producers in both the East and the West, and despite several occasions on which Ontario found itself suffering from suspensions of imports from the United States. These circumstances, which came to be widely referred to as "Canada's fuel problem," frequently gave rise to calls for a national fuel policy whereby Canadian markets would be served exclusively from Canadian supplies. This chapter will review the measures undertaken by successive Canadian governments to respond to these calls during the years between Confederation and World War I. In contrast with the period after World War I, during which the politics of coal revolved mainly around the problems of bringing Alberta coal to Ontario in the interest of national self-sufficiency, the politics of coal during Canada's first fifty years had mostly to do with the problems of bringing Nova Scotia coal to central Canada in the interest of national unity.

Nova Scotia Coal Prior to Confederation

Policies affecting the development and disposition of the fuel resources located in Canada predate Confederation itself, and the prospective disposition of Nova Scotia (and to a much lesser extent, New Brunswick) coal in a newly created Dominion of Canada was one of the issues addressed in the debates on Confederation. However, neither the early imperial policies affecting Nova Scotia coal nor the Confederation debates suggest that the coal deposits of the region were deemed to be of much significance. The early years of development in Nova Scotia were characterized by a high degree of official indifference to the coal deposits located there and, after some coal development had taken place, by persistent difficulties in finding and maintaining sufficient markets for the producers in the region. Later, the Confederation debates reveal considerable disagreement as to the desirability and feasibility of marketing Nova Scotia's coal throughout the proposed Dominion.

One of the earliest historical studies of the coal industry of Cape Breton—the location of the first production of coal in a region later to become part of Canada—describes in somewhat perplexed tones both the negligence of some of the region's earliest explorers in failing to record the unmistakable presence of coal on the island, and the reluctance (and steadfast refusal, in certain instances) of several imperial governments in later years to develop and market the resource.[2] According to this study, the first published mention of coal in the region occurred in 1672, and in 1708 a memo to the French government recommended that Cape Breton coal be exported to France. However, apart from the efforts of French and New England fishermen to carry a few tons away from outcroppings near the shore of Spanish Bay (now Sydney) no mine worthy of the name was in operation until 1720. The output of the first mine was used to meet the fuel requirements of the contingent constructing the fortress at Louisbourg and to support a small, intermittent, and clandestine trade with New England.[3]

From the destruction of Louisbourg by the British in 1758 to the formation of the General Mining Association in 1826, the mining of Cape Breton coal languished in an atmosphere of official indifference. While the imperial government did employ troops to work the coal deposits for the purpose of supplying the fuel requirements of the garrison at Halifax, it rejected several applications for leases to extract coal for commercial sale in Halifax and the other North American colonies.[4] Indeed, in December 1776, the Council at the Court of St. James declared His Majesty's royal pleasure "not at present to authorize or permit any coal mines to be opened and worked in the island of Cape Breton, and that all petitions and proposals for that purpose be dismissed from this Board."[5] The rationale for this policy, which seems to have eluded Brown, may be inferred from an earlier policy declaration issued at Whitehall in 1752:

> We are of the opinion that it would be contrary to those Rules of Policy, which this Nation has wisely observed in relation to its colonies, to bring Coals into use in America, as the use of them would naturally lead them into the Discovery of a variety of Manufactures, the raw materials of which we now receive from them, and afterward return in Manufactures.[6]

From 1784 to 1820, Cape Breton was governed as a colony separate from Nova Scotia, and during this period some leases were granted, evidently to provide revenues for the colonial administration. Production during these years was small, generally below 10,000 tons per year. Very little of this coal reached New York or Boston, a fact that Brown finds contrary to reasonable expectations, "especially since the rapidly growing cities of the United States at that time derived their whole supply of bituminous coal from Great Britain and the Virginia mines. . . ."[7] He accounts for this apparent anomaly by pointing to the bad condition in which Cape Breton coal was sent to market and blames this largely on the lack of investment in equipment and skilled labour for the operation of the mines, which in turn he attributes to the government's policy of granting only short leases at excessive rates of royalty.

Shortly after the annexation of Cape Breton by Nova Scotia, the first major investment in mining the coal of the region was undertaken by a company which was to dominate the industry for the remainder of the century. By an act of the royal prerogative executed in 1826, George IV granted a sixty-year lease of all the mines and minerals of the province of Nova Scotia to his brother, the Duke of York, apparently to help the latter out of some financial problems. This lease was then transferred for certain considerations to the General Mining Association, an English company with mining interests in Brazil and British Columbia. By virtue of obtaining this lease, purchasing an existing venture at Pictou, and negotiating yet another lease with the provincial government, the GMA became the exclusive tenant of all the mines and minerals belonging to the Crown in Nova Scotia and Cape Breton, including the most valuable Sydney mines.

According to Brown, "the chief object of the General Mining Association in opening the coal mines of Cape Breton was to establish an extensive trade with the United States, which at that time derived their principal supplies of bituminous coal from England. It was not expected that the British Provinces, where wood was abundant and cheap, would be large consumers; but it has proved otherwise."[8] He goes on to argue that, despite these earlier expectations, the growth of the Nova Scotia industry was predicated largely on the expansion of the market for coal in Montreal and Quebec and that the demand for Nova Scotia coal in the United States was not significantly affected by variations in the level, or even the complete absence, of American duties. Rather, the company's poor performance in the markets of New York and Boston resulted from the construction of a canal (opened in 1825, two years before the association began to work the Sydney mine) between the centre of the Pennsylvania anthracite coal region and the ocean port of Philadelphia.

Brown's observations on this score are interesting, as they contradict arguments commonly encountered during debates in the early years of Confederation concerning the value of the United States market to Nova Scotia coal producers and the effect on that market of the

United States tariff. The political thrust of such arguments was, of course, to impress the listener with the value to Nova Scotia of reciprocity or free trade with the United States or, to put it another way, with the cost to Nova Scotia of belonging to the Canadian Confederation. A later study, working from different sources, draws similar conclusions: the level of the American tariff was not necessarily a significant factor in determining the size of the market available to Nova Scotia coal, and the Quebec market was much more crucial than the American market to the health of the Nova Scotia coal industry. Thus,

> the initial result of the Treaty [of Reciprocity of 1854] was very slight, and a single glance at the Provincial Mines Reports shows that the real developments in the American market and the opening of new collieries began not with the Treaty, but with the civil war. That industrial stimulant brought the usual feverish activity, and the deflation and political upheavals of "Reconstruction" the inevitable collapse.[9]

Regardless of whether the United States tariff or other factors such as the reduction of transportation costs in the United States were responsible for the decline in the United States market, the fact remains that whereas in 1865 exports to the United States were of some 640,000 tons, constituting two-thirds of the total output of eastern Canadian mines, by 1875 they had fallen to 90,000 tons.[10] The Quebec market for Nova Scotia coal in Montreal was therefore becoming increasingly significant during the years immediately before and after Confederation.

Coal and Confederation

One would be hard pressed to argue that coal was a major preoccupation of the Fathers of Confederation. Coal does not even appear as a subject in the index of two prominent histories of the politics culminating in Confederation.[11] However, a major selling point of Confederation had been the potential benefits of increased trade among the contracting provinces (a matter of general urgency following the termination of reciprocity with the United States), and Nova Scotia lost little time in seeking the application of this idea to the case of coal. By 1870, owing largely to the persistence of Charles Tupper, who had led the campaign for Confederation in Nova Scotia, a coal duty of 50 cents a ton was imposed by the Macdonald government. This was the occasion of the first debate in the Canadian House of Commons on what this study has defined as Canadian fuel policy. However, the previous Confederation debates concerning the place of coal in the proposed Dominion provide a useful introduction to some of the regional conflicts of interest over fuel policies that were surfacing even at the formative stages of the new country.

A. T. Galt, finance minister of the Province of Canada, stated clearly and forcefully in the course of the Confederation debates a theme that was to be heard time and time again in subsequent debates concerning Canada's coal supply:

> Hostile tariffs have interfered with the free interchange of the products of the labour of all the colonies, and one of the greatest and most immediate benefits to be derived from their union, will spring from the breaking down of these barriers and the opening up of the markets of all the provinces to the different industries of each. In this manner we may hope to supply Newfoundland and the great fishing districts of the Gulf, with the agricultural productions of Western Canada; we may hope to obtain from Nova Scotia our supply of coal; and the manufacturing industry of Lower Canada may hope to find more extensive outlets in supplying many of those articles which are now purchased in foreign markets. . . . If we have reason to fear that one door is about to be closed to our trade [a reference to the anticipated failure to reinstate Reciprocity with the United States] it is the duty of the House to endeavour to open another, to provide against a coming evil of the kind feared by timely expansion in another direction; to seek by free trade with our own fellow-colonists for a continued and uninterrupted commerce which will not be liable to be disturbed at the capricious will of any foreign country.[12]

D'Arcy McGee, Minister of Agriculture and Statistics, and the representative of an urban riding in Lower Canada, put even greater stress on the potential benefits to be derived from access to the fuel resources of Nova Scotia. Having presented the assembly with a table summing up the balance sheet of the proposed union in statistical and financial terms, he proceeded to make reference to "one special source of wealth to be found in the Maritime Provinces . . . the important article of coal." He went on to point out that wood was giving way to coal as the major source of household heating in parts of Canada; that there had been great suffering among the poor of Montreal from the high prices of fuel; that there was no coal in Canada; and that Canada has a five-month winter, generally very cold. He then referred to the coal resources of the maritimes, "to whose mines Confederation would give us free and untaxed access forever." He concluded by rebutting the anti-unionist argument that Confederation would mean the loss to Nova Scotia of the New England market on the grounds that "we consume in this country as much fuel per annum as they do in all New England; and, therefore, that we offer them a market under the union equal to that which these theorizers want to persuade their followers they would lose."[13]

Not all the members of the assembly, however, were as enthusiastic about the benefits to be derived from union and the interprovincial trade it was supposed to bring about. H. G. Joly, representing Lotbiniere, submitted to the assembly a prophetic assessment of the trade prospects of the proposed union:

> The Gulf Provinces possess timber, coal and fisheries; our own two great articles of export are timber and wheat. With regard to timber, the Gulf Provinces have no more need of ours than we of theirs. As to coal we import from the Gulf Provinces. When this supply becomes insufficient to meet our growing wants, it will be necessary to look somewhere for a supply of coal. If the Lower Provinces can furnish it to us at cheaper rates than we can get it in the United States, we shall buy it from them. Upper Canada will probably get its coal from the Pennsylvania mines, which are in direct communication with Lake Erie . . .[14]

Similarly, Adolphe Gagnon, from Charlevoix, questioned the supposed benefits of the Inter-Colonial Railway and the "great commerce" which it was claimed would be opened up with the maritime provinces, on the simple grounds that "those provinces have nothing to exchange with us . . . They have nothing but coal which we do not possess, and that is not transported by railway."[15]

The potential benefit of the coal trade that was supposed to result from the proposed union was not, however, the only controversial point about coal raised during these debates. Some concern (and, indeed, no little confusion) was expressed concerning the article of Confederation that would permit Nova Scotia to impose an export duty on minerals, including coal, traded by the province. Some members of the assembly were sympathetic to Nova Scotia's desire to obtain a revenue from their coal and likened the export duty to timber dues in other provinces, while others objected that there was in fact an important difference between a royalty and export duty, one that favoured Nova Scotia in that it would pay nothing on the coal consumed in the province but would charge a duty on the goal it exported to the rest of Canada.[16] The power of Nova Scotia to impose such a duty was, however, retained in the provisions ultimately passed by the Canadian assembly.[17]

The Canadian debates on Confederation reveal that coal was hardly a vital issue in the central provinces and was not unanimously regarded as even an important one. Moreover, what some members did say about Confederation was not uniformly favourable to Nova Scotia coal, while what they said about Nova Scotia coal was not necessarily favourable to Confederation. To the advocates of Confederation, however, this coal had at least indirect significance as one of the important elements in the interprovincial trade which, they argued, would compensate for the loss of reciprocity with the United States and would help to forge bonds of unity for the new nation. But even these indirect benefits were contingent upon the ease with which interprovincial trade in coal could be established and sustained; and this, in turn, was dimly perceived to depend on the degree to which the central Canadian buyers would be willing to pay a premium on the Nova Scotia product. The first real indication of the willingness of these potential buyers, and an indication of the importance to Nova Scotia of interprovincial trade in coal, may be found in the debate that soon followed in the new Canadian Parliament over the imposition of a duty against imported coal.

The First Coal Tariff, 1870

The first move by a government of Canada to encourage interprovincial trade in fuels was tentative, tumultuous, and short-lived. It is hard to distinguish clearly the principal forces that led to the early demise of this experiment, partly because the government was itself equivocal about the objectives the tariff was intended to serve, and partly because criticism of the tariff came from a coalition of those who opposed protection in principle, those who opposed the protection of coal in particular, and those who opposed the legislation in question because of the protection it simultaneously extended to other items, such as wheat and flour. To make precise interpretation of this debate even less certain, the government itself was at one stage prepared to withdraw the bill.

The policy at issue was a proposed duty of 50 cents per ton on coal and coke introduced to the House as part of the budget of Sir Francis Hincks, the minister of finance in the Macdonald government. When first challenged to justify the proposed tariff, Hincks stressed the idea that it was, "on the whole," a strictly revenue tariff and not "one that will expose us to any strong charge on the score of being protectionist."[18] On the other hand, he also referred favourably to a statement by his predecessor, who had warned the Americans in the previous year that coal could not continue to enter freely into Canada from the United States while the Americans maintained a heavy duty against imports of Canadian coal, thus suggesting that the tariff was meant either as a retaliatory measure or as a token with which to negotiate the removal of the American duty.

Within three weeks, Hincks was announcing the government's determination to abandon the duties on coal and wheat, but not on flour.[19] The coal duty apparently had fallen victim to petitions from Montreal manufacturing interests, strident denunciations from the opposite side of the House, and scepticism even from the government's own ranks. The difficulty with the revised tariff, however, was the ire it would raise in Nova Scotia, since flour was to remain under protection without the compensation of protection for coal. As one Halifax member put it, to the audible concurrence of Charles Tupper, he had supported the duty on flour on the supposition that the finance minister would place a duty on coal. "The Government had taken the members from Nova Scotia by surprise." He wondered whether any Nova Scotia member could support the bill as it now stood. "If so," he said, "they would deserve the execration of their constituents and he hoped they would get it." These sentiments evidently prevailed over the opponents of the original scheme, for the next day, Hincks announced that the government was to return to the tariff as it was first introduced.[20] The motion required to bring this about was finally passed on April 28, 1870. A petition from Toronto against the duty was already before the House by May 3.[21] The duty against coal was repealed by the Commons about a year later.[22]

Even a cursory reading of this debate would seem to support the view that the tariff on coal in particular—if not protection in general—had less going for it than against it from the very beginning.[23] The spectre of the government's failure to achieve reciprocity with the United States was present even during the most strident speeches in support of the tariff, and the argument that the tariff would encourage interprovincial trade in lieu of North American trade was imbued with a tone of resignation. Indeed, at several points the government seemed to intimate that one virtue of the Canadian tariff might be its usefulness as a lever in bargaining with the Americans for a reduction in theirs. Even the Nova Scotia supporters implied conditional rather than absolute endorsement of the tariff, since their arguments frequently stressed that coal should be protected as compensation for the protection on wheat and flour. In brief, there was little positive enthusiasm for Macdonald's vision of a comprehensive National Policy, in whose service he had advocated the 1870 tariff, even on the government side of the House.[24]

Among the points raised in opposition to the coal tariff (apart from equivocal opposition to protection as such), three stand out. The first and most telling was the argument that the 50-cent tariff would be insufficient to place Nova Scotia coal at an advantage in the Ontario market, and thus represented little more than a tax on coal consumption in that province, with no perceptible gain to Nova Scotia producers. This argument appears to have been empirically sound since, as discussed below, transportation subsidies were required, in addition to a tariff, before Nova Scotia coal was to make even a minor dent in the Ontario market. A second point was that the tariff applied to both bituminous (soft industrial) coal, which Nova Scotia produced, and to anthracite (hard domestic or home-heating) coal, which Nova Scotia did not produce. Thus, the householders of Canada were also being taxed on their consumption of domestic fuel, again with no perceptible gain to Nova Scotia producers. Finally, there was the straightforward argument that the interest of Quebec and Ontario consumers in acquiring coal from the cheapest possible source should prevail over the interest of Nova Scotian producers in obtaining larger markets. Representatives from Central Canada tended to stress that injury was being done to nine-tenths of the people of the country for the benefit of one-tenth, and that even Nova Scotians were suffering as a consequence of the government's protectionist stance, especially with respect to wheat and flour, so that the only clear Nova Scotian beneficiaries from tariffs on coal and other commodities were the coal operators themselves.

Given the lukewarm enthusiasm of many supporters of the tariff, together with the stridency of many others in opposition, Forsey's comment that the persistence of Charles Tupper was largely responsible for the adoption of the tariff seems a reasonable assessment of the debate.[25] Charles Tupper, in addition to being one of the statesmen of Confederation, represented Cumberland, one of the three coal-producing counties of Nova Scotia.[26]

Coal and the National Policy

With a few exceptions (to be discussed shortly), little noteworthy parliamentary debate on the coal industry occurred between the repeal of the coal duty in 1871 and its reintroduction as part of Macdonald's National Policy in 1879. However, a brief debate did arise in 1876 on a motion by Thomas McKay of Cape Breton concerning the coal trade between Canada and the United States, and in the course of this debate Tupper again urged the adoption of a national policy. It is worth noting that the year 1876 marked a five-year low in Nova Scotian coal production, and the lowest point in sales to the United States since at least 1850.[27] McKay's argument for his motion contained three points of interest.[28] First, he sought action by the government to protect the coal industry of Nova Scotia by "either obtaining the removal of the duties imposed by the United States on our coals or levying an equal duty upon American coal." Second, he placed emphasis on the danger of allowing Ontario to remain "entirely dependent upon a foreign country for every ton of coal consumed in it," and submitted that "there is only one place on which Quebec and Ontario can depend with any degree of certainty for their coal supply, and that is Nova Scotia." Last, he raised the issue of national unity: "Unless we have certain trade interests protected; unless the East can trade with the West and exchange commodities I do not think this confederation can hold together."

Tupper reiterated most of these arguments. Both he and McKay seemed to have had an eye not only to erecting a tariff that would make Nova Scotia coal more competitive in Montreal, but to establishing both tariffs and transportation subsidies sufficient to make it competitive in Ontario. In the interest of promoting this trade between Nova Scotia and Ontario, he advocated subsidies for lines of steamers transporting Ontario flour to the Maritimes and bringing back coal as return freight, "without increasing its cost to the people of Ontario."[29] With this plan, Tupper argued, not only would the industries of the country be fostered, but, "you would protect Ontario against the great danger of leaving her dependent upon the monopolies in the United States which, after crushing out the coal industries of our own country, would demand their own prices."[30]

The arguments against the plan outlined by McKay and Tupper suggested that Ontarians were simply not prepared to make the sacrifice they deemed it would require, namely an increase of about $2 per ton in the price of coal in Ontario. As one member put it, it would be

> impossible to put enough duty on to force us to deal with Nova Scotia. We should be very happy to deal with her, as I think it is out of our line to trade with our American friends; but if we can save $2 a ton it is our duty and our interest to do it.[31]

In the following year, McKay again presented the House with an opportunity to debate the coal issue, beginning this time with "the petition of Henry Mitchell, and others, coal owners, and others interested in the coal trade and shipping interests of the Dominion, praying that the duties imposed by the United States on Canadian coal, may be neutralized by an equal bounty on all coal sent to the United States; or that a duty of 50 cents per ton be imposed on all coal imported into the Dominion."[32] This petition was ruled out of order as involving a public charge, and McKay had to settle for a motion to refer the petition to a select committee.[33]

McKay's motion was debated in terms resembling those of previous debates on the coal tariff question. He again stressed the dependence of Ontario and Quebec on foreign supplies of coal; the languishing state of the coal industry in Nova Scotia, owing to the falling-off of the market in the United States; and the injustice of offering protection and other supports to numerous other Canadian industries while showing what seemed to him to be "determination" on the part of the present and previous governments to stamp the coal industry out of existence.[34] Support for McKay's motion was forthcoming not only from the predictable quarter of members representing coal-producing regions, but also from some Conservatives or Liberal-Conservatives from central Canada who wished to see the coal issue resolved within a general policy of protection. Thus, Joseph Ouimet, a Liberal Progressive from Laval, invited McKay (of the Reform Party) to "join the great party which was waving the flag of protection, and help on the movement to bring about a national policy of protection to all industries." J. B. Robinson of Toronto sympathized: "Nova Scotia was more or less forced into the Confederation on the understanding that the question of coal should receive the attention of the Dominion." He pointed out that $2 million was taken out of Ontario every year for the benefit of Ohio and the Western States, instead of "our fellow countrymen in the maritimes," and stressed further the interests of Ontario shipowners, shipbuilders, and manufacturers in the expansion of the interprovincial trade that could be fostered through some protection to the coal industry.

Opposition to this motion came largely, though not exclusively, from Ontario and Quebec, and generally reflected scepticism that a duty in the order of 50 cents per ton would produce any movement of coal from Nova Scotia to Ontario, even though it might, as one member observed, increase Canadian consumption of Nova Scotia coal as far west as Montreal.[35] Most of those opposing the motion felt that a duty of at least $1 to $2 a ton would be required to open the Ontario market, and this represented an intolerable burden on the manufacturers and homeowners of central Canada. Thus, William McGregor, a Liberal from Essex, sardonically supported the motion on the ground that the proposed inquiry would only confirm the impracticality of the whole idea: "When the facts were brought out by the committee, the House would, no doubt, see that Ontario could not purchase coal from Nova Scotia."[36] Perhaps the most notable voice in opposition was that of Alfred G. Jones, who, although from Halifax, roundly condemned the idea of a duty on coal and of protection in general, on the ground that its advocates generally wished to see protection extended only to those commodities in which they or their constituents had an interest.[37] He also made the potentially helpful observation, albeit to little avail, that the petitioners themselves sought consideration of the bounty or duty on imported coal "in the interests, not of protection, but of free trade." In other words, Jones was astute enough to notice that the coal duty was sought as a means of expanding the Canadian market for Nova Scotia's coal only if an expansion of the American market through reciprocal trade arrangements with the United States was not attainable.

The motion to refer the matter to a committee was eventually agreed to, and a report was produced the same year. Read in conjunction with the minutes of evidence heard by the select committee, the report shows that the solution they envisaged to the coal problem can be fairly described as the National Policy in microcosm.[38] In drawing their report to a close, the committee stated broadly the direction of their hopes and, implicitly, their recommendations: "The great importance of encouraging the closest commercial relations between the provinces of the Dominion, induces the strongest hope in the minds of your Committee that the efforts of those at present engaged in endeavoring to promote inter-provincial trade, may be successful." This seems to be a reference to several representatives of coal companies in Nova Scotia who had expressed confidence that they could compete with American exporters in such places as Toronto and Montreal, provided return freights were available for the journey back from Ontario ports to Nova Scotia, thus reducing considerably the freight charges borne by Nova Scotia coal being shipped to Ontario. Without such an arrangement, the committee estimated that the differential in favour of U.S. coal at Toronto would be about $1.50 per ton. However, several witnesses had testified that, with sufficient two-way traffic between Ontario and Nova Scotia, the latter's coal could compete quite favourably with the American product. For example, during the proceedings one Nova Scotia mine manager stressed that on the higher grades of coal, no duty would be required to make Nova Scotia coal competitive with American coal in Ontario even "as far as London, provided we could get down freights; it is contingent on down freights."

The importance of return freights to the interprovincial trade in coal dovetailed nicely with the interest of Ontario farmers and manufacturers in expanding markets in the Maritime provinces. Both the Hamilton and Toronto boards of trade forwarded identical resolutions to the committee objecting to a customs duty upon the importation of coal simply by itself but approving a customs duty on coal as part of "a revision of the tariff in the interest of the general industries of the country . . ."[39]

From the record of the committee's deliberations, several assumptions seem to have stood behind their all-but-explicit recommendation of a duty on coal. (It will be remembered that the committee would have been out of order in making an explicit call for a duty.) First, a general policy of protection would enhance the competitive position of both Nova Scotia coal in Ontario and Ontario foodstuffs and manufactures in Nova Scotia. The movement of coal would be further promoted by the reduction in freight charges made possible by the two-way flow of goods. Moreover, as demand for Nova Scotia coal grew, unit costs of production in the mines would decline, permitting even greater sales. Last, with greater prosperity in Nova Scotia, the corresponding increase in demand for Ontario produce in that province would maintain the necessary volume of return freight.[40]

Not all the evidence placed by witnesses before the committee supported these assumptions or the duty they were called upon to justify. For example, the Chamber of Commerce of London was evidently dubious about both the effects of the duty on the price of coal in their city and the benefits to be obtained by their region from enhanced interprovincial trade. At any rate, they petitioned the committee to reject the idea of a duty on either anthracite or bituminous coal.[41] The Londoners may have been aware of the stance taken by the owner of a local gasworks, who later testified that "we could not touch Nova Scotia coal" even after a duty of $1 per ton had been imposed against the American article.

The committee came to grips with several other controversial points concerning the coal trade, such as the possible effect of the duty on British and American coal imports into Montreal and the effect on the market potential of Nova Scotia coal in New England in the event that the United States government reacted to a Canadian tariff by removing the American one. But the key issue, and the most controversial point in the later debate in the House, was the likelihood that a 50-cent tariff would in fact work to open the Ontario market to Canadian coal in the first place. Nevertheless, the proceedings and report of the standing committee seem to have had some impact on the Macdonald government, for duties of 50 cents against coke and anthracite and bituminous coal were announced by Tilley in the general schedule of tariffs commonly regarded as part of the National Policy.[42] When its turn came up for individual discussion about a month later, the coal duty received fairly brief consideration, and was passed without division.[43] On this occasion Tilley defended the tariff on the ground that "under this proposition, Nova Scotia coal would be brought to the western part of the Dominion." When Alexander Mackenzie greeted this assertion with the question, "How far west?", Tilley replied with a statement that hints at the possibility of an "Ottawa Valley Line" for coal. In his own words, he

> had no doubt a large portion of Nova Scotian coal would reach Toronto. There was no doubt whatever that it would be supplied to all the small towns this side of Belleville, certainly to Montreal, and Montreal would be able to supply a portion as far as Toronto.

Mackenzie then proceeded to denounce the tariff as an unjust tax that discriminated against one part of the country, apparently for the benefit of another—"although he denied that it would materially benefit any part."

Tupper followed Mackenzie with a lengthy and spirited defence of the duty in terms that appear to owe much to his earlier presence at the proceedings of the select committee. Most of his comments were aimed at persuading the House of the validity of the assumptions held by the select committee concerning the potential effect of the duty on interprovincial trade and the benefit to be derived from that trade by both Nova Scotia and the central provinces. However, it is noteworthy, and somewhat surprising, that before arguing on these grounds, he took pains to suggest that the Canadian tariff could result in the elimination of the American one, thus reviving the prosperity of the Nova Scotia industry. He believed that

> the effect of this policy would be to give free coal to both the United States and Canada at a very early date. The whole object of the duty on the part of the United States would fall to the ground the moment Canada adopted a policy similar to theirs. Then the natural result would follow that the coal mines of Nova Scotia would supply the Atlantic States, and the coal mines of the United States would supply coals to Toronto and the western portions of Canada.

It would appear from this that the contribution of increased inter-provincial trade to the unity of the new Dominion was not deemed to be intrinsically beneficial but desirable only to the extent that the prosperity contingent upon Canadian–American trade could not be realized because of *American* protectionism. In this respect, Tupper's position was completely in line with the earlier petition from the coal producers of Nova Scotia, who had insisted that the intent of the duty they were pressing for was to induce the United States to agree to reciprocal free-trade arrangements. It was not to be the last occasion on which Canadian fuel producers expressed interest in a national policy only as compensation for the failure to achieve satisfactory continental marketing arrangements, or "swaps," as they would come to be called in the era of oil and gas.

Again, opposition to the duty, when it was not based simply on objections in principle to protection of any kind, tended to be based on scepticism concerning the likely effectiveness of the tariff in opening the Ontario market to Nova Scotia, the argument being that unless it did so, the duty amounted in effect to a tax on Ontario manufacturers and householders. (There was similar opposition to the tariff on anthracite coal, which Nova Scotia did not produce.) For instance, Brown referred to the duty as an experiment to be tried at the expense of Ontario. "They might put the duty at $1.50 a ton, but the people of Ontario would still have to import American coal . . . He did not think they could bring coal from Nova Scotia in large quantities unless by water. It would not do to return flour and wheat in coal vessels, and they would have to find some other kind of freight for the return trip."[44]

The criticisms of the coal tariff were to no avail, and coal producers, among countless other industries, are perhaps most appropriately regarded as beneficiaries of support for the National Policy in general. In any case, protection of the industry was solidly entrenched, as the duty on bituminous coal was never again to fall below fifty cents, although anthracite was placed on the free list as early as 1887.[45]

In the event, it appears that the duty did produce the desired result of increasing the demand for Nova Scotia coal in at least part of central Canada, although there is some uncertainty as to how much of the subsequent increase in the production of coal in Nova Scotia ought to be attributed to the effects of the coal duty alone and how much to the more general industrial expansion which came about as a result of the opening of the western territories, the rapid construction of railways, and other key elements of the National Policy. For example, Nova Scotia coal producers also benefited from tariffs on imports of iron, steel, and steel products, owing to the fact that they supplied the fuel requirements of the growing Nova Scotia steel industry.[46] At any rate, "the returns for 1880 compared with 1878 show an increased production of 262,000 tons, and increased sales to Quebec of more than 155,000 tons."[47] However, as Forsey goes on to point out, the Toronto and southwestern Ontario market was scarcely touched. In 1880 the duty was raised to 60 cents, but even this had no discernible effect on Nova Scotia's sales beyond Eastern Ontario, reinforcing the impression that the coal tariff was designed primarily to prevent American coal from encroaching on Canadian markets any further east than the Ottawa Valley, and was not meant to open the Ontario market to Canadian coal. (This impression is further reinforced by the fact that coal shipments from Nova Scotia to Montreal were conducted mainly by water, but would have had to proceed further west by rail at much higher cost.) Nevertheless, "in 1880 production again exceeded one million tons and continued a steady growth, almost doubling between 1885 and 1891 and again between 1891 and 1901."[48] Exports grew over the same period by only about 600,000 tons. By 1913, production reached its all-time high of eight million tons. The politics, therefore, of the period between the adoption of the National Policy and the advent of World War I might be expected to show signs of complacency and satisfaction, at least on the part of producers.

Coal Debates, 1880–1913

There were few debates concerning coal during the thirty-odd years following the adoption of the National Policy. There were occasional speeches from individual members who complained that the tariff against U.S. imports did either too little or too much, depending on whether the member in question represented Nova Scotia or Ontario, respectively. As early as 1884, for example, a member rose to inquire, and implicitly to complain, about the limited extent to which Nova Scotia coal was actually penetrating the markets of central Canada. Specifically, he complained that according to the results of inquiries he had made, the only purchaser of Nova Scotia coal in the city of Ottawa was the government itself, which used it to heat the Parliament buildings.[49] He speculated that the failure of the miners of Nova Scotia to secure larger markets in Canada might be a result of excessive freight charges on coal coming over the Intercolonial. At any rate, he received a reply assuring him that "coal from Nova Scotia has been delivered very largely as far west as the town of Brockville." This seemed to end the matter. In 1894, there were calls in Parliament from three members for a reduction or even elimination of the duty against American coal, which was seen as an onerous tax on the manufacturing industries of Ontario and as an added burden upon the country's railways, "struggling now with the restricted conditions of traffic and everything else," as one of the members put it.[50] The same member summed up his view; "We are glad to see the miners of Nova Scotia having a portion of this trade, but it is too bad that the entire trade of the country must be hampered in order to benefit them."

One speech made during this period bore a close resemblance to earlier ones advocating a national policy. In March 1909, D. D. McKenzie, a Liberal from Cape Breton, placed figures before the House to demonstrate that "American coal is gaining ground along the river St. Lawrence in the province of Quebec, to the exclusion of our own coal."[51] (Other sources for production in Nova Scotia show that 1909 marked a decline of about 800,000 tons, following many years of steady increase. Production increased again, however, in the following year.[52]) McKenzie appealed for measures to be taken so as to "hold the markets of Canada for Canadian coal."[53]

While clearly a call for a national policy on coal, McKenzie's argument departs on several points from earlier arguments in favour of such a policy; some of these points seem almost to be premonitions of the debate on Canadian fuel policy that was to take place in the 1920s. For instance, McKenzie argued for the familiar idea of bringing Nova Scotia coal as far west as Toronto, but also suggested the possibility of supplying the rest of central Canada from the coal fields of the Prairies. The desirability of such a policy was defended in part on the ground of reducing Canada's dependence on a foreign source of fuel, and McKenzie's fellow members were invited to "imagine what would happen if our fuel supply in the United States were stopped in mid-winter." If the people of Ontario were to give sufficient thought to the position in which they would find themselves in such an eventuality, he warned, they would realize that "the wise policy on our part is to develop our own coal resources and coal trade and not depend upon the whims of any outside country." He also stressed the familiar theme of maintaining the prosperity of Nova Scotia as a necessary means to sustain the "great trade which we furnish to the people of Canada", having earlier pointed out that probably 75 percent of the goods consumed in the province come from Ontario and Quebec. However, another note is struck at several points in McKenzie's speech, introducing a new tone to the debate on a national coal policy, one that would come one day to predominate: in addition to keeping an eye on the interests of investors in the country, Parliament "should also keep in mind the men, women, and children who are dependent upon this particular industry in every part of Canada . . . We have in Nova Scotia 103,000 people who are depending for their livelihood upon coal cutting and mining." He urged, therefore, that

> some measures should be taken to enable these working men to have employ-
> ment every day and to earn sufficient money to support themselves and their
> families in comfort. I regret that this winter they have not been kept in con-
> tinuous employment because of this displacement of Canadian coal in the
> home market by American coal.

As we shall see below, the 1946 Royal Commission on Coal, one of the last of the numerous committees and commissions to inquire into the Canadian coal situation, was to read more like a document on social policy than one on energy policy. It seems fair to conclude, following Fowke, that coal has held an appropriate place in the National Policy, old *and* new.[54]

Conclusion

The federal government's interest in matters involving coal during the years reviewed in this chapter was not restricted to the place of coal in the National Policy or to a national policy respecting coal itself. How- ever, these concerns were certainly predominant during the debates on coal policy that took place during the first fifty years of Confederation. To most of its advocates during those years, a national coal policy was regarded (and was similarly understood, though not desired, by its op- ponents) as an attempt to meet all Canadian demand for coal from Cana- dian sources. In the arguments for this policy emphasis was placed not simply on improving the health of the industry in Nova Scotia and the prosperity of that province, but especially on the two-way trade that would grow up between separate regions of Canada, creating bonds of mutual interest that would reinforce the union. The insecurity of foreign supplies was not ignored, but it was not central to most of the arguments made.

Against these considerations were weighed the simple facts of geography and economics, combined with an unwillingness on the part of consumers in central Canada—especially Ontario—to pay the extra cost involved in overcoming the constraints imposed by both types of factors. Nova Scotia coal was truly competitive only as far west as water transport would carry it, and even in these markets it was not immune to competition from American coal. It was clear that considerably

greater tariff protection or some additional form of transportation subsidy would be necessary to extend the market for Nova Scotia coal as far
west as Toronto. The amount of Nova Scotia coal consumed west of the
Ottawa Valley was generally limited to less than 10 percent of the
bituminous coal consumed there.[55] Estimates of the assistance that
would have been necessary to increase that figure ranged substantially
during the debates from a tariff of over two dollars a ton down to
nothing, provided the proper two-way trade could be developed between Ontario and Nova Scotia. In any case, the measures that would
have expanded the Canadian market for Canadian coal were strongly,
though not unanimously, resisted in Ontario and they were never sufficiently applied. This failure did not result in the dire consequences for
Canadian unity predicted by such early advocates of the national policy
as Charles Tupper; but this was probably because the industry did pros-
per despite the lack of markets in Ontario, owing initially to the demand
for iron and steel that was occasioned by that other component of the
National Policy, the construction of railroads, and, later, to the gradual
growth in demand for Nova Scotia coal and steel as Quebec industrialized and the economies of the maritime provinces grew.[56]

However, World War I and the years shortly thereafter were to produce two conditions that would alter this combination of relative complacency in Nova Scotia about markets and indifference in central
Canada to Canadian supplies: First, during the war, the Montreal market
was lost to Nova Scotia producers. Second, in the winter of 1922, coal
deliveries from the United States to Canada were suspended because of a
long strike of the mineworkers there. The result of these developments
was an intense and prolonged campaign for a new version of a national
coal policy.

3   Coal and National Independence, 1919–1946

IF, GENERALLY speaking, the national politics of coal for the first fifty years of Confederation centred on the theme of fostering national unity by increasing the volume of trade between the provinces of Canada, then the national politics of coal during the next thirty years may be said to have centred on the theme of fostering the independence of the nation as a whole by replacing foreign sources of fuel with Canadian supplies. National unity gave way to national independence as the primary rationale and justification for a national coal policy. Similarly, and not entirely coincidentally, the major impetus for the new national policy came not as it had earlier from the peripheral, producing provinces, but from the central, consuming provinces, especially Ontario. The most distinguished advocate of a national policy for coal in the cause of national unity had been Charles Tupper, who represented a coal-producing region come on hard times after losing markets in the United States. The leading advocate of a national policy for coal in the new cause of national self-sufficiency was Thomas Church of North York, a former mayor of Toronto, who campaigned tirelessly for the adoption of an idea which, for a time, was widely supported in his province: to gain access to Alberta coal once coal imports from the United States had proven unreliable. The degree of success that Church's cause was able to achieve is the subject of the present chapter. A few background points concerning the experience of the country with regard to fuel supplies during World War I and the years immediately following the war will place this campaign in perspective.

The War and Fuel Control

The most salient point in the country's experience with fuels during World War I was the loss of the Montreal market to Nova Scotia coal, since part of the politics of the postwar period was to be concerned with federal measures aimed at helping Nova Scotian producers regain this market. Nova Scotia lost the Montreal market as a result of the diversion of ships from interprovincial coal trade to other uses occasioned by the war.[1] In the meantime, Quebec had substituted increased imports of American coal for Nova Scotia supplies, and no real shortage of coal was felt in Canada until the winter of 1916–17, when the expansion of industrial activity spurred by the war began to push demand for coal in the United States and Canada beyond the productive capacity of the continent. By midsummer of 1917, prices in the United States had reached unprecedented heights, prompting President Woodrow Wilson to order them adjusted and fixed. In view of this and more general anxieties about the nation's coal supply, the Dominion government appointed a fuel controller on June 11, 1917.

The wartime activities of the fuel controller are of less interest here than are a few of the observations he made in submitting his final report to the government in 1919.[2] It is fairly clear from the report that the main concerns and major activities of the controller and his administration involved, on the one hand, allocating supplies available to Canada among various regions and users and, on the other hand, negotiating with the United States Fuel Administration to ensure that Canada did not suffer unduly (i.e., disproportionately) as scarce United States supplies were allocated among previously established markets for American coal. At this, by his own estimation, he was quite successful, considering that "during the crisis just past, Canada was treated almost, if not absolutely, as an integral part of the union."[3]

By the time he wrote his report, however, the controller had different problems on his mind, since the urgency of meeting the demand for coal had disappeared with the cessation of munitions production at the war's end.[4] Nevertheless, the wartime experience had apparently underlined the point that Canada needed to become more efficient in its consumption of coal and, most important, had to find a means to reduce the dependence of central Canada upon a single form and a single source of coal for its home heating needs:

> The consumer of hard coal in central Canada—Ontario and Quebec—might as well realize that the Anthracite producing district is practically confined to the State of Pennsylvania; that any disturbance in any section of the field will immediately affect the whole district, which in turn affects the entire market. Furthermore, 72 per cent of the total anthracite mined is produced by eight powerful corporations, and if any of them should at any time see fit to become arbitrary, it would cause, at the least, serious inconvenience to the market . . .
>
> Owing to the adverse conditions surrounding anthracite supplies, central Canada is far too cold a country to depend on it alone.[5]

He stressed the importance to the consumers of central Canada of acquiring the capacity to turn to the use of soft coal during periods of emergency. This had apparently proven difficult during the war. Of further interest, especially in the light of developments to be reviewed shortly, was the controller's assessment that one of the obstacles to the increased use of soft (bituminous) coal, in addition to consumer inertia, was the fact that anthracite producers *and dealers* are naturally opposed to an invasion of their market, through the use of soft coal. The fields producing this latter fuel are so widely distributed, and there is such an abundance of it, that its market *cannot be controlled by trade agencies*, as is possible with anthracite supplies.[6]

This potential resistance on the part of consumers and coal dealers in central Canada was soon to be activated over the question of expanding the use of Canadian bituminous coal in their markets.

Complementing the fuel controller's concern over the vulnerability of central Canadians to disruptions in the supply of imported coal were two other concerns. One was the contribution that increased interprovincial trade in coal could potentially make to the task of maintaining a favourable trade balance for the country:

> No field offers more scope for national or individual effort toward that end than does thrift in the production and consumption of coal. The coal operators, both in the East and in the West, should be encouraged as far as possible to make the most of the domestic markets, and thereby decrease the amount of foreign coal imported.[7]

The other concern, that of increasing the markets available to Canadian producers, dovetailed neatly with this: "It is impossible to forecast the extent to which the maritime producers will regain the Quebec market; and to what extent the Alberta and Saskatchewan operators will be able to hold the Manitoba market." There was, moreover, a proportion of the home market that Canadian producers had never been able to enter, owing to the distance from the coal fields; and the extent to which this home market could in future be served by the Canadian product was regarded as "a problem that requires careful consideration."[8] It is something of an understatement to say that this problem did receive some consideration during the next decade.

Coal in the West

It would be misleading to suggest that the national coal debate of the 1920s was concerned only with the problem of serving the Ontario market with Alberta coal, but it seems fair to say that this issue was the predominant and most controversial one. It would be only slightly more misleading to suggest that the whole of this controversy revolved around rates on the rail transportation of coal from west to east, but there is a historical precedent for doing just that. An Alberta premier once attempted to sum the problem up for the benefit of Mackenzie King: "Alberta has the coal. Ontario needs the coal. It is purely a transportation problem . . ."[9] The dubious word in this remark is "purely"; but the basic point seems fair enough, for one of the most contentious, most ambiguous, and most thoroughly studied aspects of the Canadian coal problem during the 1920s and 1930s was the issue of the cost of transporting Alberta coal to markets in Ontario.

It will be remembered that Nova Scotia's coal production reached a peak of some eight million tons in 1913. By that same year, a considerable coal industry had grown up in the western provinces as well. For example, production in 1913 was, in round terms, over four million tons in Alberta and two and a half million tons in British Columbia. (Insignificant quantities of lignite coal were produced in Saskatchewan before the war.) The main impetus for development of the coal industry of Alberta had been the development of the transcontinental railways. However, there is evidence that it is not quite true to say, as the Royal Commission on Coal later did in its report on the Canadian coal industry, that "production in Alberta started with the railway construction by the Canadian Pacific Railway Company."[10] The commissioners do not appear to have encountered the following inscription on a cairn in the Galt Gardens in Lethbridge, Alberta:

> In 1872, on the western bank of the Oldman River Nicholas Sheran opened the first coal mine in Alberta. He broke his own trails, found his own markets, and hauled coal by bull-team 200 miles to Ft. Benton, Montana and other distant points.[11]

Exports, however, were largely inconsequential to later producers in Alberta, with sales to railways and householders accounting for almost 80 percent of total sales, the remainder being taken up by industries and coke and gas plants.[12] However, production in British Columbia, which until 1898 was confined to Vancouver Island, was directed largely to markets in California, other Pacific Coast states, and Pacific areas as far-flung as Alaska, Chile, the Hawaiian Islands, and Japan.[13] Production in the Crow's Nest region of British Columbia was stimulated in 1898 by the construction of the CPR, and from that year forward nearly all of the growth in coal production in British Columbia was based on demand for this coal from the railways, the metallurgical industries of the province, and the northwest United States.[14]

By 1913 there was a substantial coal industry in the two most westerly provinces. Three features of the western industry were of some consequence to government action and political debates with respect to Canada's fuel problem during the 1920s. First, there was the partial (about 25 percent) reliance of West Coast producers on export markets, particularly the American Pacific Northwest. Because of the fear that the American government would match any Canadian tariff against United States coal with an equal tariff against American imports of Canadian coal, British Columbia producers by and large opposed tariffs designed to protect Nova Scotia coal from American competition in central Canada or to promote the eastward movement of Alberta coal. This opposition prevented a unanimous posture on the part of the Canadian coal industry and coal-producing provinces on the issue of protection. The second salient feature was the apparent inability of Alberta producers to compete with American imports even as far west as Winnipeg, whose household market was served with American anthracite coal, and whose industrial market was served with American bituminous coal.[15] One of the success stories of the 1920s was Alberta's winning over of the Winnipeg market by means of an aggressive marketing campaign and some federal assistance in the form of rail subventions; but this story also tells how impractical it must have been to promote the sale of Alberta coal in competition with American imports another thousand miles or so further east in Ontario.[16] The third feature was the chronic over-capacity of the western industry and the constant pressure from producers for government measures to assist them in reaching wider markets. All these features of the western coal industry played a part in national developments during the late 1920s and early 1930s concerning the interprovincial movement of Canadian coal and in the extended political debate over the desirability and feasibility of expanding the Canadian consumption of Canadian coal.

Steps Toward a National Coal Policy

The question of what assistance, if any, should be granted by the federal government to bring Alberta and Nova Scotia coal into Ontario was first given serious consideration by the Special Committee of the House of Commons on the Future Fuel Supply of Canada during the 1921 session. The most important, or at any rate the first, recommendation of this committee—the proceedings of which will be reviewed below—was the appointment of an officer with powers to handle fuel supply emergencies and generally to study and report on Canada's fuel situation. This recommendation was given greater urgency by two events in the following year. An extended coal strike in the United States during the winter of 1922–23 prompted the Minister of Mines to recommend, in November 1922, the creation of a Dominion Fuel Board to carry out the recommendations of the House committee as well as those contained in the final report of the fuel controller, some of which have been referred to above. This was accomplished by means of an order-in-council in a matter of weeks, and the new board was given the mandate to study Canada's fuel problems "in view of the ultimate necessity of substituting other fuels for anthracite coal for domestic heating purposes in Central Canada."[17]

As if to emphasize the urgency of this matter, the United States Fuel

Control Board proved unable to authorize adequate deliveries of anthracite coal to Canada, and there was a serious shortage of fuel for Canadian households during the winter of 1922–23. The Dominion Fuel Board warned that this situation could become permanent, owing not only to labour troubles but also to a more general shortage of anthracite looming in the United States.[18] On March 8, 1923, a special committee of the Senate was appointed to consider these matters. It reported on June 25, drawing the attention of the government to the importance of a reduction of freight rates on Canadian coal shipments and recommending that the DFB be empowered to cooperate with transportation companies and other interests to improve the transportation, handling, and storage of Canadian coal. Thereafter, the idea of assisting the movement of Alberta coal into Ontario received considerable attention from a variety of parliamentary committees and government agencies and, as we shall see below, spawned one of the most protracted energy debates in the nation's history.

The federal government seemed to bide its time throughout all this, despite the plethora of recommendations for action from various committees, commissions, and boards, although it did proceed to increase substantially the subvention of the westward movement of Nova Scotia coal from $65,600.58 in 1928 to $205,220.16 in 1929.[19] It required, however, a change of government to bring about significant action in aid of Canadian consumption of Canadian coal. In 1931, R. B. Bennett instituted a system of freight subsidies that were to raise the total federal assistance to the coal industry to over $1,200,000 by 1933. He also raised the tariff on imported coal from 50 to 75 cents, and made several other minor changes.[20]

Bennett's measures had an observable impact on the share of the Canadian market available to Nova Scotia producers. For example, whereas Nova Scotia bituminous coal made up only 16 percent of all such coal consumed in Canada in 1926, it made up 27 percent of such consumption by 1939.[21] Montreal and Ottawa took the majority of this coal, however. Thus, even when the figure for Ontario includes Nova Scotia coal consumed in eastern Ontario, no more than 13 percent of the Ontario market was served by Nova Scotia coal, compared with 63 percent of the Quebec market in the same year. Alberta was a different story, at least with respect to central Canadian markets. The combined effect of the tariff and the freight subvention (in Alberta's case, a flat $2.50 per ton) was insufficient to facilitate any significant movement of that province's coal into Ontario. The amount of Alberta coal shipped to Ontario under this subvention varied between a low of about 30,000 tons in 1932 to high of about 92,000 tons in 1939. Even the high figure, of course, represents a miniscule fraction of the Ontario market in that year of over twelve million tons. However, it is clear that the combined effect of federal policies did work to increase the competitiveness of Alberta and British Columbia coal in Manitoba and the Head-of-the-Lakes region, so that deliveries of about 650,000 tons of western coal were assisted annually to this market during the middle and late 1930s.[22]

In summary, it can be said that between 1922 and 1939, some government action was taken in the direction of increasing the Canadian consumption of Canadian coal, but only on a limited scale, so that the Ontario market in particular remained largely dependent on American supplies. This is not surprising, perhaps, given the prohibitive cost involved in moving sufficient volumes of Canadian—particularly Albertan —coal into this market, a cost which the Royal Commission on Coal was later to estimate at no less than $100,000,000 a year. But it does seem a little surprising in the light of the sustained and intense campaign for more positive action that took place during the mid-1920s. Throughout this period the government was under pressure from its parliamentary opposition, the government of Alberta, and local Ontario politicians (to name the most prominent sources) to establish a national fuel policy, and the terms in which the debate over this policy was conducted are part of a revealing chapter in the history of self-sufficiency as a goal of Canadian energy policies.

The Campaign for a National Fuel Policy, 1921-1946

The campaign for policies to achieve Canadian self-sufficiency in coal came from both the producing and the consuming provinces. The parliamentary campaign was led by Thomas L. Church—lawyer, former mayor of Toronto, "ardent protectionist," prominent ally of Sir Adam Beck in the struggle for public ownership of hydroelectric power in Ontario, director of the Comsumers Gas Company and several public utilities, and Liberal-Conservative member for Toronto North.[23] Church did not initiate parliamentary discussion of the crisis of 1922-23. In fact, two major energy debates took place before he moved, in 1924, that "in the opinion of this house, the time has arrived for Canada to have a national policy in relation to its coal supply and that no part of Canada should be left dependent on United States [sic] for such supply."[24] He did, however, initiate the first call for government action in that direction, whereas earlier debates had been concerned only with more thorough study of Canada's fuel situation. There can be no doubt that the primary impetus for parliamentary agitation on this issue was the fear in central Canada that the United States might no longer be a reliable source of coal, especially home heating coal (anthracite). These fears were sharply exacerbated (though not necessarily caused) by two strikes in United States coal fields, disrupting deliveries to Canadian markets.

Movement in the direction of a national coal policy began in March 1920, when a motion was placed before the House citing the "acute" fuel problem in the country and calling for a full parliamentary discussion of the rising price of "necessaries." The originator of this motion, J. H. Burnham (Peterborough West), made reference to two documents that expressed concern about the unreliability of American coal and Canada's heavy dependence on that coal, and further called for the use of Canadian coal throughout all of Canada. (These two documents, incidentally, suggest what seems at first glance to be an unlikely alliance between the manufacturers of Ontario and the coal producers of Alberta.) The first document was a letter informing Burnham that the Canadian Manufacturers' Association (CMA) had struck a special committee "to investigate the possibility of using Canadian coal for the whole of Canada."[25] The letter further reminded Burnham of the coal strike in the United States in the previous year, a strike that "at one time promised to create a more serious situation for Canada than during any part of the war period." Of special concern was the fact that Canada had no way of "adjusting the strike one way or the other," while Ontario was "solely dependent upon that part of the United States affected by the strike for our supply of coal." Moreover, coal imports might be vulnerable to other interruptions, in retaliation for Canada's failure to export sufficient quantities of pulp and paper to the United States. The letter finally reminded Burnham of the enormous coal reserves in Alberta and concluded that "the solution to our fuel situation would appear . . . to be a matter of transportation, and the Government now owns two transcontinental railways."

The second document, which had accompanied the letter to Burnham, was a copy of a report of the Coal Section of the Edmonton Board of Trade. This report cited at considerable length the CMA resolution encouraging greater use of Canadian coal in Canada. It pointed out Alberta's potential capacity to serve Canada's needs, argued that the Canadian coal market was the key to the expansion and increased efficiency of the Alberta industry, and went on to record its own resolution

> favouring a most thorough inquiry into the cost of transportation of coal in railway cars to Ontario, as well as by lake and rail; and pledging outselves to assist the Canadian Manufacturers' Association in every way possible to secure necessary information to lay before our Federal Government on this matter with a view to enlarging the markets for Canadian coal in Canada.

The sections of the CMA resolution cited in the Coal Section's report are worth reproducing in full, partly because this resolution appears to have been the spur to public debate and eventual government action on the matter of bringing Alberta coal to Ontario, and partly because they present a fairly succinct statement of the terms in which the public debate on this issue was to be conducted with varying intensity for an entire decade. Burnham read the CMA resolution into the record as follows:

1. Whereas Ontario is not possessed of any coal deposits, and is therefore dependent upon a supply of its fuel coming from a foreign country over which the Ontario consumers and the Government of Canada have no control; and,

Whereas the industrial life of Ontario is largely dependent upon the good will of the American people for its supply of coal; and,

Whereas there is always the possibility of the United States being so situated that its whole supply of coal might be required to insure its own protection; and

2. Whereas the Government railways in Canada, particularly the Canadian National, and the Grand Trunk Pacific which apparently the Government contemplates operating, are yearly faced with heavy deficits which arise through lack of sufficient traffic to assure a profitable conduct of their business; and

3. Whereas the Province of Alberta is possessed of unlimited quantities of all varieties of coal suitable for Ontario's needs; and,

4. Whereas Alberta is a large market for Ontario manufacturers, and that this market will best be maintained and consolidated through the correct methods of commerce being applied, namely, by selling to Ontario as well as buying from them; and

5. Whereas Canada is largely indebted to the United States through the development of our own resources and industries, with the result that the time seems to be approaching when we will only be able to purchase from the United States for cash.

Therefore, it is in the best interest of Canada that the Dominion Government, and the Government of the Province of Alberta should cooperate to help bring about the desired condition of making Canada independent of any foreign country in its fuel supply, and thus retain for the development of Canada the means of exchange which comes about through using our own coal . . .[26]

Having put these documents before the House, Burnham went on to observe that numerous public bodies and private persons agreed with the assessment of Canada's fuel situation contained in the resolutions and were beginning to realize "that Ontario is in danger of a coal famine, and all the horrors incident to a coal famine, and that this is a disaster that may be imminent." He concluded by insisting that it was a "prime necessity" to consider the question now, since "heretofore people simply threw up their hands, and impotently asked, 'What are we going to do?' "[27]

If the terms of the CMA resolution were indicative of the arguments to be heard for years afterward in favour of a national coal policy, the government's response to Burnham's motion was equally indicative of the argument brought time and again against it; It would be too expensive. In the course of the brief debate following Burnham's speech, Arthur Meighen, Minister of the Interior, allowed that the quantity of coal available in Canada was not an issue, but that the problem lay in its accessibility: "The coal of the Dominion lies mainly in the extreme provinces." The difficulty facing the coal industry of Canada was that the market "within transportation range" of the coal fields was too limited to allow production on a sufficiently large scale to achieve competitive costs, and the cost of transporting this coal to centres of heavy coal consumption was prohibitive.[28] Putting his finger upon what was to emerge as the heart of the matter, Meighen went on:

> I could not give the details, but I can assure the hon. member that all these subjects have been most carefully inquired into by those who would like to develop these mines and to reach the very market the hon. member has in mind. They are the most anxious people of all. There is no doubt in the world that the Edmonton Board of Trade would be delighted beyond measure if some means could be found that would reduce the cost of transportation, possibly at the expense of the Canadian National railways, to such a figure as would bring about a larger area of consumption for the coal of Alberta. We all would like the same thing, but these things must be put on a common sense business basis, and the efforts of the Mines Department are devoted to the finding of ways and means for putting the business on that basis in order that it may be made a success on its own merits, and not at the expense of failure of some other national enterprise.

The issue turned, as it would continue to do for years, on what a "common sense business basis" was, or should be taken to be. In other words, a lengthy controversy developed over several matters of both fact and judgment: What does it cost to transport a ton of coal from Alberta to Toronto? Given the cost, what freight rate is appropriate? If that rate is too high to make Alberta coal competitive in Ontario, what is to be done? Make the Ontario users pay higher prices? Make the railways haul coal at a loss? Subsidize the freight rate out of general tax revenues? Reduce the return to producers? Or a combination of these? Clearly Meighen meant to suggest by his reference to "a common sense business basis" that the government was not justified in doing anything, and successive governments, Conservative and Liberal, were to adopt much the same attitude. But the public would not have it. Burnham's motion was merely an early indication that pressures were mounting in favour of a national fuel policy, pressures that were to peak during the ensuing five years.

A year later, in March 1921, continued difficulties with the price and availability of coal in Ontario prompted Michael Steele (Conservative,

South Perth) to introduce a motion in Parliament to appoint a special committee of the House of Commons to inquire into all matters pertaining to the future fuel supply of Canada.[29] Steele was primarily concerned with the costs and dangers of allowing Ontario to depend entirely upon the U.S. for its coal supplies. He documented the fact that the retail price of imported coal was increasing rapidly (from $15.50 per ton to $22 per ton in only three or four months of the previous year) owing to its scarcity in Canada, and pointed to the probability that the people of the United States would soon insist on the suspension of anthracite exports within the foreseeable future. In short, Steele argued that Ontario needed to give greater attention to its future fuel supply and that Canada should increasingly look to its own coal resources for a solution. He did not advocate bringing Alberta coal to Ontario, and he seemed to feel that the answer lay in developing a transportation policy that would permit Nova Scotia to supply the Ontario market. In the course of the generally favourable debate which followed, several ideas were suggested to meet Steele's objectives, such as using idle grain cars during the summer to carry western coal to Ontario, and improving the St. Lawrence route so that coal boats could work directly between Nova Scotia and Ontario lake ports. Support came from members from both the producing provinces and Ontario, and from both the Conservative and Liberal parties. Arthur Meighen, now prime minister, supported the motion with the reservation that the country could not hope for some time to carry Alberta coal farther east than the head of the Great Lakes. In both the Nova Scotia and Alberta cases, in Meighen's view, the key problem was the cost of transportation:

> You cannot carry coal for nothing. You may make the rate nothing, but that does not mean you carry it for nothing; it means only that you carry it at the cost of the country: and if you carry it at the cost of the carriage of other commodities of course it comes out of the people. There is a certain inherent cost in the carriage of coal and no amount of bookkeeping is going to reduce it.[30]

This observation seemed to close the debate, and the motion was passed.

It can be seen from these two motions and the debates on them that concern about Canada's fuel situation, particularly the country's dependence upon the United States for its coal supply, was growing prior to 1923, as was support for a solution in the form of a national fuel policy. It was further becoming clear that a national fuel policy came down essentially to the matter of freight rates on shipments of coal from Nova Scotia and Alberta to Ontario. The winter of 1922-23 gave an enormous boost to the growing campaign for a national policy and drew even more attention to the problem of bringing Alberta coal to Ontario.

Again, a United States coal strike during the summer of 1922 was the prime mover. Described as "the greatest coal strike in the history of the civilized world," it resulted in severe hardship in Ontario the following winter.[31] To make matters even worse, there was an atmosphere of anxiety about the capacity and willingness of the Americans to continue to supply home-heating coal to Canada. What had always been seen as a virtue by some—national self-sufficiency in fuels—was now seen as a necessity by many more, and the result was increased support for a national fuel policy. Perhaps the most important indication of this new level of support was the success of T. L. Church's motion in the House of Commons endorsing the idea of a national fuel policy:

> That, in the opinion of this house, the time has arrived for Canada to have a National Policy in relation to its fuel supply and that no part of Canada should be left dependent on a United States coal supply.
>
> That further in the opinion of this House, the government should immediately consider the initiation of an all-British and Canadian coal supply and that such a policy is both a social and economic necessity and in the best interests of the future of Canada.

The idea embodied in Church's motion was to be debated off and on for a quarter century, by which time the idea of a national coal policy would be rendered academic with the displacement of coal by oil and natural gas as principal fuel sources. The two sides to this debate were divided primarily by different views of the cost of transporting coal from Canada's producing provinces to Ontario, and by different views as to the desirability and feasibility of substituting Canadian grades of bituminous coal for anthracite coal for household use. Beyond these issues, it is interesting to observe that considerable support for a national coal policy existed for a time in the major consuming province of Ontario, owing to a pessimistic view of the future of American supplies of anthracite combined with an optimistic view of the feasibility (at acceptable cost) of transporting Canadian coal to Ontario.

Given that the political controversy over the issue of moving Canadian coal supplies to Ontario involved so many conflicting claims with respect to such seemingly objective matters as the cost of hauling coal by train, it is unfortunate that the truth of this matter is inherently difficult to ascertain.[32] Nevertheless, the Board of Railway Commissioners looked carefully into such matters in the late twenties and early thirties and arrived at cost figures in the order of $8 to $10 per ton. They concluded that the 1923 published rate of about $12.50 per ton was reasonable and only slightly in excess of the operating cost. These figures should help to keep the political debate on these matters during the twenties in some perspective.[33] But despite the possible ambiguities regarding the cost of transporting coal, there is no question that the substitution of Alberta coal for imported coal involved an enormous increase in the aggregate cost of meeting Ontario's fuel requirements, even according to figures cited by those advocating policies to bring this movement about. For example, the report of the special committee of 1923 had this to say:

> . . . your Committee has no hesitation in recommending that every possible effort should be made by those in authority to encourage the public to obtain their supplies of coal or other fuel from Canadian sources. The fact that we imported for consumption last year 13,017,025 tons of coal at an approximate cost of $61,112,428 from the United States and other countries should impress everybody with the necessity of utilizing our own fuel resources to the fullest extent.[34]

The enthusiastic tone of this assessment is difficult to understand, given the fact that the aggregate import figures cited amounted to no more than a unit cost of $4.70 per ton, or about one-third of the cost of *transporting* Alberta coal to Toronto at prevailing freight rates. Even the most optimistic estimates of the cost of moving Alberta coal to Ontario would have meant that the Canadian fuel bill would be at least double the $61 million quoted above. Nevertheless, to the advocates of a national policy, from whatever region, the villains of the piece were the railways (for insisting upon exorbitant freight rates on coal shipments) and the federal government (for letting them get away with it). Thus, Church himself, in speaking to his motion, quoted a letter from "a gentleman in Alberta" to the effect that "a 2,000-ton train of coal from an average point in Alberta to the average point in Ontario, mileage 2,200, would leave a profit of almost 25 per cent at $6 per ton," even making allowance for the returned empty cars.[35]

After hearing several voices in support of the national fuel policy he was advocating, Church eventually heard from the Minister of National Defence (the Honourable G. P. Graham). Graham did not wish to prevent the issue from going before a Committee of the House, but in the course of his remarks he took issue with Church on the question of freight costs. Not only were operating costs up in general, owing largely to higher wages, but the notion of transporting Alberta coal to Ontario was particularly unrealistic in the light of what economists would term the problem of surplus capacity: "The trouble with freight originating in the West is that it is nearly all eastward . . . so that when we bring coal east we add to the traffic without having any return traffic to carry back."[36] The consequence of this, to which Graham had already referred, was that part of the cost of hauling a ton of coal from west to east consisted of the charge for hauling the empty car back again. In any case, he added, "Everything else being equal, coal can be hauled from mines 500 miles away and brought to your door more cheaply than coal hauled from a point 1,800 miles away." As an alternative solution,

Graham mentioned the idea of increasing the use of Welsh coal in Ontario, and conceded that there might be some sense in increasing Canadian consumption of Canadian coal by hauling it in the summer months when eastbound traffic was less in other products. He also conceded the advisability of converting from anthracite to bituminous coal, noting that a first-class bituminous coal, properly fired, would dispense as much heat as anthracite coal.

All in all, Church received no outright opposition to his motion, despite some scepticism, and the motion was passed without division. So the Commons began yet another inquiry into Canada's fuel problem. (The issue had been examined only two years earlier by a special committee of the House, and the previous committee, while it had heard testimony on the prospect of bringing Alberta coal to Ontario, had recommended no action be taken in that direction.[37]) Reporting that "many expert witnesses were called and some considerable volume of evidence taken" on the issue of transportation costs, the House committee considering Church's motion found that the views expressed on freight rates were "very divergent and inconsistent" and recommended that the Minister of Mines immediately call a conference of government representatives and interested parties to examine the matter more closely. The committee further recommended that the government undertake an independent study "to ascertain the actual cost of carrying coal from eastern and western points to Central Canada," and added that "we believe that our National Railway should carry fuel at cost in this crisis, and your Committee suggests that the rates quoted are not cost rates but much higher."[38] (In contrast, a special committee of the Senate, which was simultaneously investigating Canada's fuel situation, referred to rates of between $9 and $12.40 per ton, and reported its "strong inclination" toward the view that Alberta and Nova Scotia coal could not be expected to overcome the handicap it faced in the Ontario market "unless the railway companies are prepared to transport coal at less than cost."[39])

The upshot of the House committee's 1923 report was yet another motion from Church calling for a national policy in relation to Canada's fuel supply and calling upon the government to bring this about by means of, first, tariff protection for coal mined and coked "under the British flag" and, second, preferential tolls on Canadian railways for Canadian coal.[40] After again providing evidence from several sources of the unreliability of coal supplies from the United States, Church went to work on the railways, and particularly the Board of Railway Commissioners, for their failure to establish tolls "in consonance with the actual costs and conditions." Church quoted from an application to the board from the government of Ontario, alleging that the rates filed by the railways for the transportation of coal were excessive and were preventing the marketing of Canadian coal in the province, and demanding that the board direct the railways to give an account of themselves. As for Sir Henry Thornton, president of the Canadian National Railways, Church complained that he

> has gone up and down this country quoting one tariff rate here, and another tariff rate there. I should like to know if Sir Henry Thornton is the court of last resort on this question, bearing in mind that we have a Railway Commission in this country that is empowered under the Railway Act to bring into force rates that are just and reasonable in the interests of the people.

What the country needed, according to Church, was the kind of leadership shown by President Coolidge in the United States, who apparently "called the officials of the railway companies and the mine owners before him . . . and declared that unless they reduced their rates he would submit a measure to congress to control the situation."

Most of the extended debate that followed was in support of Church, and most of it focused on freight rates, although tariffs were also discussed. J. J. Hughes (Liberal, Prince Edward Island) argued in favour of a swapping arrangement, whereby Alberta coal would be sold in adjacent markets in the United States in exchange for United States supplies in Ontario. But more centrally at odds with Church was Arthur Meighen, now leader of the opposition. Meighen opposed the idea of increasing the tariff against coal imports from the United States on the ground that this might raise the cost of production in central Canada to a point where they could no longer be internationally competitive.[41] He also opposed the principle of putting coal in a special category "and compelling our railways to levy very low rates in accordance with that category." In other words, Meighen was resisting the inclination of most advocates of the national coal policy to assume that the cost of its implementation should come out of the railways' profits. Finally, Meighen defended the independence of the Board of Railway Commissioners:

> I do not see any consistency, any common sense in appointing a Railway Commission with specific powers, with specific duties, and then after appointing it and empowering it to exercise such duties immediately stepping in and exercising some of these duties ourselves. If we intend to perform this function, for goodness sake let us remove the duplication and wipe the railway commission off the slate.

(This was not the last occasion on which a part of the controversy surrounding Canadian energy policy has had to do with the independence or otherwise of a regulatory agency with jurisdiction over the transportation of fuels.)

Nevertheless, Meighen was in sympathy with the objective of decreasing Canada's reliance on American coal. He observed that the attention given to the problem in recent years had intensified because of "events that have seemed to threaten our supply altogether." Moreover, Canada's economic difficulties were aggravated "because of the withdrawal from domestic circulation of so much of our money for the purchase of American coal."[42] Even more crucial, however, was the disadvantage (not to say indignity) of allowing the United States such a powerful lever over Canada in negotiations and disputes over other matters:

> I do not like to be in a position with respect to the Chicago drainage canal or any other question where I am liable to be answered, by implication, if not expressly: Unless you yield, look out for your coal. If that answer lies in the mouths of American statesmen, whether they express it or not, it is going to be reflected in the results of any discussion we have on . . . any . . . subject of controversy.

In sum, there is the implicit suggestion in Meighen's speech that reasonable measures taken toward a national coal policy are justified not simply because they reduce by themselves the magnitude of Canada's dependency upon the United States but also because they help to preserve the health of the Canadian coal industry, whose very existence places some limit upon the American capacity to exploit that dependency.

> I said that we could even now meet an emergency, but at great cost. If we were denied American coal, though certainly American coal interests would be very strongly against it, if we were for any reason in that position, we could bring coal from Alberta, we could bring coal from Nova Scotia.

The government's position was not far from that of the leader of the opposition. The minister of railways, for example, appeared to share Meighen's view that Parliament should not be interfering with freight rates, given that these were under the jurisdiction of an independent board.[43] Charles Stewart, minister of mines and the apparent spokesman for the government in energy matters, also appeared to support the principle of Church's motion, but tended to emphasize the difficulties the country faced in bringing it about, most of which resolved themselves "into a question of transportation and transportation alone." If it had not been for the water route between Nova Scotia and Montreal, he pointed out, it would be impossible for Nova Scotia coal to compete with American coal in that market. As for the rail rate on western coal, it was the "back haul empty that caused the high rate on one way haulage," a problem also affecting rail shipments of Nova Scotia coal to Central Canada.[44]

Stewart made one other point that deserves to be underlined and might have been given more serious attention by the advocates of a national coal policy: he observed that Canadian coal not only faced strong competition at the time from American anthracite in the home heating market, but would continue in future to face strong competition from American bituminous coal coming from that country's vast deposits which lay within three hundred miles of Ontario markets. The notion of a national coal policy in concrete (or technical) terms would have meant, in part, the displacement of about 4.5 million tons of American anthracite coal per year by an equal amount of Canadian coal—coke made from Nova Scotia bituminous coal, and Alberta's somewhat unusual "domestic" coal. What the advocates of this policy seemed to overlook was that American bituminous coal could also be coked for use in Canadian homes, and its locational advantage would be as telling in future as American anthracite's was in the present.

But these considerations were not sufficient to silence the advocates of a national policy. As already indicated, the matter was debated off and on at greater or lesser length for another twenty years, largely because the policy was assuming the proportions of a national cause and was receiving considerable attention and support from experts, academics, newspapers, local officials, and the general public in central Canada. For example, the *Canadian Annual Review of Public Affairs, 1922*, reported that Arthur V. White, consulting engieer to the Ontario Hydro-Electric Commission, had been persistently presenting to the public his views on the Canadian energy situation, which included recommended measures for the independence of Canada in its fuel supplies along with the suggestion that Canada should barter its hydroelectricity with the United States for an assured coal supply.[45] The same source quoted editorials in four different central Canadian newspapers, all concerned in one way or another with Canada's dependence on outside sources of fuel. A year later, the *Annual Review* of 1923 reported that "keen interest" was shown by easterners in demonstrations of the use of Albertan coal and that "there was no difficulty in disposing of the limited amount of coal available."[46] The report went on to quote an editorial from the *Montreal Herald* to the effect that the idea of substituting Alberta coal for United States imports "has greatly stirred the imagination of our people" and was, moreover, one which would not be left to rest with Sir Henry Thornton and his $9 rail rate on coal shipments. The report noted that representatives of the Alberta mines had appeared before the Toronto Central Council of Ratepayers, "which endorsed their proposals and resolved for action."

There is further evidence that consumers throughout Ontario eventually came to endorse using Alberta coal in their province and supported the growing campaign, engineered by Alberta's trade commissioner, Howard Stuchbury, to pressure the federal government into taking more positive action to assist them in doing so. They were joined in support of this campaign by various local boards of trade and by the Canadian Manufacturers' Association. (It will be remembered that Burnham's first motion on a national fuel policy, introduced in 1920, made reference to a resolution by the CMA supporting such a principle.) The files of the office of the premier of Alberta for the years 1923 to 1926 contain numerous references to the interest in Alberta coal that had been found throughout Ontario.[47] As early as August 1923, for example, a letter was sent from the town clerk of Dundas, Ontario, to the premier of Alberta, expressing support for sales of Alberta coal in Ontario and suggesting that freight rates could be blamed for the difficulties standing in the way of this trade.[48] In March 1924, similar letters of support were sent from the city of London and the city of Owen Sound. By September 1926, the Union of Canadian Municipalities was writing to inform the premier that their convention held in London, Ontario, had passed a resolution calling for "some definite action to assist and encourage the transportation of coal from the Provinces of Alberta and Nova Scotia to the other provinces."[49]

Similarly, the Alberta Trade Commissioner had been able to report to his chief in December 1923 that the Associated Boards of Trade of Western Ontario "were most enthusiastic and passed a very strong resolution" in favour of bringing Alberta coal to Ontario.[50] Shortly thereafter, he was able to report further, from Toronto, that "the government here have had the most enthusiastic reports as to the efficiency of our coals, in fact the whole of the Province, except the American coal dealers, are sold on Alberta coal."[51] Moreover, "every member of the Ontario cabinet is behind this campaign, and none of them rarely address [sic] a meeting that they do not advocate a satisfactory rate for Alberta coal." The upshot of all this, apparently, was considerable pressure on the federal government to intervene in some manner to promote the movement of Alberta coal to Ontario. One Toronto businessman, for example, wrote to congratulate the Alberta premier on his success in bringing the issue to the point "where it has almost upset the whole country, and the biggest men in the country now tremble." There was, he added, evidence that the issue could hurt the King government in the next election.[52]

Both of these claims could well have been true, but neither the "biggest men in the country," if they were indeed trembling, nor Mackenzie King, if he was indeed threatened politically, were mobilized beyond anxiety to the stage of effective action. The files of the Alberta premier's office contain many references to the lassitude, not to say the outright obstruction, of the federal government on the matter of assisting the movement of Alberta coal. Certainly the early responses of the government on the matter of the Alberta coal shipments show every sign of indifference. For instance, King's private secretary was able to reply rather unenthusiastically to a letter from Premier Greenfield to the prime minister which contained a copy of a resolution passed by the legislative assembly of Alberta "requesting the cooperation of the Federal Government in an effort to secure a freight rate on coal that will enable Alberta coal to enter the Ontario market in competition with American coal."[53] The reply, which apparently was not followed up, informed Greenfield merely that the prime minister had taken the occasion to bring the matter to the attention of his colleagues, for their consideration.[54] Meanwhile, C. A. Magrath, chairman of the Federal Advisory Fuel Committee, was informing Greenfield of his feeling that "it is impracticable to bring Alberta fuel to Ontario on account of the great distance it has to come."[55] He added the suggestion that perhaps the province itself could subsidize the railways in making such hauls, "by giving them the coal mineral rights to some fixed tonnage."[56]

By 1925, the response of the federal government and several of its agencies appeared no more satisfactory in the eyes of the principal figures in Alberta. Stuchbury complained to the premier that "the attitude of the Dominion Fuel Board from its inception has been antagonistic to Alberta coals . . ."[57] A little later, in an interesting observation in the light of the parliamentary debates taking place at the time, Stuchbury complained that delay on the part of the federal government was solely responsible for the failure of a planned experimental shipment of Alberta coal to Ontario. "It is patent in the face of correspondence," he wrote to the Premier, "that the blame cannot be laid at the doors of the C.N.R. but might properly be laid in its entirety on the Minister of Mines and the Ottawa Cabinet."[58] It is interesting that the Alberta actors in this drama were not inclined, as their Ontario allies were, to look to the railways as either the villains of the piece or to mandatory rate reductions as the solution to the problem. The evidence for this goes well beyond Stuchbury's complaint. Upon hearing that Premier Ferguson of Ontario had applied to the Board of Railway Commissioners for a reduction on the rate of shipments of Nova Scotia and Alberta coal, Greenfield sent a telegram to Ferguson complaining about the lack of prior consultation with Alberta, insisting that such an application would be "extremely prejudicial to our cause, as the rate necessary in such a case would upset the whole freight rate structure," and insisting instead that the

> rate if granted must be voluntary on part of railways and should be made as a special arrangement with Governments to meet special national situation and not as published rate which would create precedent for future applications for freight reductions.[59]

The federal government did not remain adamant on the issue of transporting Canadian coal to the Ontario market. A trial shipment of Alberta coal was finally arranged, and appeared to meet its one objective of dismissing prejudices concerning the quality of Alberta coal, if not its other objective of ending the controversy over the "real cost" of shipping a ton of Alberta coal to Ontario. In 1927 a rate was set well below that which the railways had claimed to be necessary for that purpose, and the government made up the difference. Tariffs were put in place as added protection for Canadian coal against the American product, the manufacture of coke was underwitten, and a variety of other interventions were tried. But throughout the 1920s and 1930s, Ontario remained dependent on imports of American coal, and the cries for a national fuel policy continued unabated.

One would search in vain for arguments that were new and of any significance in the ensuing debates on the desirability and practicability of a national coal policy. Between 1924 and 1946, references to coal in the House of Commons are too numerous to mention, much less relate, but the number of extended debates on a national policy is smaller.[60] Especially in the 1920s, the focus of these debates continued to be the issue of freight rates and the reluctance of the Canadian railways and the Board of Railway Commissioners to set a rate on the transportation of coal that would permit Nova Scotia and (particularly) Alberta coal to compete in the Ontario market. One of the frequent allegations, made more or less explicitly at different times and by different critics, was simply that the railways were insisting on a rate which was "excessive," by which we assume that the critics meant the railways' quoted rate of $9 to $12 per ton would result in higher than ordinary profits. But one wonders if even proof in these matters would have made a substantial difference. Who can really argue with the proposition that, all other things (such as production costs) being equal (which they were, roughly), coal that travels 2,000 miles to a market will be more expensive than coal that travels 500 or 800 miles to the same market? The important question was not whether a "real" difference existed between the cost of Canadian and American coal in the Ontario market, but rather whether national self-sufficiency in fuels was worth the extra cost. Answers to this question depended largely on, first, how urgently Canadian coal appeared to be needed and, second, by whom the difference in cost was to be paid: producers, railways, the general taxpayer, or the Ontario manufacturer and homeowner. Finally, the willingness to pay on the part of those who would be called upon to pay would naturally be affected by the total cost per annum of maintaining self-sufficiency. To make things worse, all of these matters were ultimately arbitrary; that is to say, they were amenable to no "objective" resolution or even definition. Those who, for whatever reasons, favoured the goal of a national policy tended to minimize costs or attempt to find some other party to absorb them. Those who did not favour it tended to exaggerate the costs or refuse to pay them.

In the end, those who favoured a national fuel policy failed, although not completely. The federal government did act to expand the market for Canadian coal beyond what it would have otherwise been. But the actions of the federal government fell far short of those required to make Canada self-sufficient in fuels. In the words of one advocate of the national policy, this must be seen as a victory of the "natural economy" over a "willed economy."[61] As we have seen, the will to create a national fuel market was widespread, and was not at this time restricted to fuel-production regions. Indeed, the parliamentary campaign was led by a Torontonian, and the policy was supported by manufacturers, municipalities, and newspapers in central Canada. But the intensity of this support seemed to vary with the urgency, or perceived urgency, of Ontario's fuel needs. The less reliable American coal appeared, the greater the demand from Ontario to gain access to Canadian supplies. While the several emergencies brought about by interruptions in United States supplies were a sign to some that Canada should establish permanent means of meeting its own fuel needs, it would appear, as Arthur Meighen suggested, that only a sustained and severe shortage of fuels would have brought about the sacrifices necessary to supply Ontario from Alberta and Nova Scotia continuously.

The exact size and nature of these possible sacrifices were always a controversial matter, but there is at least some indication of their order of magnitude. Charles Stewart, while minister of the interior, put the extra cost of replacing American imports with Canadian coal at $56 million in 1927.[62] This figure agrees with one arrived at in an academic analysis published in 1931: "The difference in expenditure on coal between an all-Canadian supply and a supply under free market conditions would represent annually a sum of perhaps forty to sixty millions of dollars."[63] Assuming that Stewart's figure for 1927 was reasonably accurate, it is worth noting that the total expenditure of the Dominion government in the same year was slightly over $350 million, and that a national coal policy would have meant roughly a 15 percent increase in public expenditure in that year.[64] It seems fair to conclude that an expenditure of this magnitude might be considered a small price to pay to escape an actual fuel crisis, but a very large annual premium indeed for an insurance policy against one. Thus, despite the protest on the part of some national policy advocates that "Alberta and Nova Scotia cannot continue indefinitely as mere emergency coal-bins," there is at least a reasonable case to be made that it was in the national interest that they remain precisely that.[65] It was quite clearly in the central Canadian interest, at any rate.

The campaign for a national fuel policy probably failed because the dangers of dependence upon United States imports gradually declined in salience after 1925, the last year prior to World War II in which Canada faced a shortage of fuels. (Ironically, the shortage on this occasion was the result, in part, of a strike in Cape Breton mines.) Even if the prospects for long-term supplies of American anthracite were somewhat gloomy, by the thirties it was understood that the coking of bituminous coal would increasingly serve the household market in Ontario, and the supply of bituminous coal in adjacent states of the United States for this and industrial uses appeared without limit. More important, coal was beginning to face increasing competition from oil. By 1946, while the Royal Commission on Coal (the Carroll Commission) was beset with demands for a national fuel policy, the calls for such a policy were almost exclusively voiced by petitioners from the producing provinces.[66]

The picture of the Canadian coal industry as presented to the royal commission by producers and the producing provinces immediately following World War II was not very encouraging. During the war, American coal had made inroads into markets in Quebec and eastern Ontario previously held by Nova Scotia, and into markets in northern Ontario and Manitoba previously held by Alberta.[67] Meanwhile, the industrial demand brought about by the war had resulted in an increase in productive capacity in all the producing regions. In addition, Canadian coal producers, especially those in British Columbia, were losing markets to fuel oil, and several briefs pleaded not only for increased subventions on the transportation of Canadian coal to Canadian markets, but also for an increase in the tariff on imported oil.[68] The almost unanimous call from producing regions was for a national fuel policy, by which was meant a policy aimed to increase Canadian consumption of Canadian coal, primarily by means of government assistance for the transportation of Canadian coal.

As familiar as this call may seem by this time, it is possible to detect a slight shift in emphasis in the arguments used to justify such a policy. Compared with those advocating a national fuel policy in the 1920s and 1930s, the parties recommending such a policy to the royal commission in 1945 said less about the virtues of complete self-reliance in fuels and more about the need for equity in the economic relations between central Canada, whose manufacturing industries benefited from tariff protection at the expense of all Canadians, and the coal-producing provinces, whose coal producers needed public assistance in order to survive.[69]

Meanwhile, parties in central Canada argued that no special action should be taken to restrict the use of American coal in Canada. For example, the government of Ontario referred in its brief to "the generous treatment accorded the people of this Province by the United States

Government in the allotment of coal to meet the needs of our people during the war," and suggested that "any proposal that would have the effect of placing restrictions on the marketing of American coal in Ontario should be closely considered from this point of view."[70] Similar sentiments were expressed by coal distributors and dealers in Ontario and Montreal.[71]

Without discussing in detail the eventual findings and recommendations of the Carroll Commission, it seems fair to conclude that the commission's thinking ran more along the lines represented by parties in central Canada than along those represented by parties in the producing provinces. "Independence may be physically possible," the report of the commission concluded, "but it is too impractical to merit further attention. Central Canada must therefore continue to rely mainly on United States sources of supply."[72] The commission saw some merit in extending assistance to the coal industry beyond that provided by the prevailing tariff, and saw the preferable form of assistance as some form of transportation subvention. Such a policy was "only fair," in view of the advantages accruing to Ontario and Quebec under Canada's fiscal policy. The primary justification for such assistance was not to provide for Canadian energy requirements under normal conditions, but to meet social and economic needs in the producing provinces:

> At least 100,000 people are dependent, directly or indirectly, on Nova Scotia coal production; without aid additional to the present tariff the industry will be unable to support that number, with resulting social and economic dislocation. Some dislocation in the coal areas of Canada must also be anticipated if additional assistance is not provided. . .

Despite the additional comment that "the maintenance of a reasonable level of production in Canadian mines may be of special value during periods of emergency,"[73] the proposed subventions were not seen primarily as a means to ensure self-sufficiency in coal, or any other objective relating to Canadian energy supply as such, but were seen as an instrument of Canadian welfare policy. This seems to have set a kind of precedent. Over thirty years later, approval of the Alaska Highway pipeline would be primarily justified, not as a means of expanding Canada's natural gas supplies, but as a means of stimulating the national economy, particularly the flagging manufacturing industries of Ontario.

4 Nationalism Versus Continentalism: Oil and Natural Gas Pipelines, 1949–1958

WE HAVE SEEN that during the age of coal (that is, during the period in which coal was the primary source of fuel for the country) the notion of a national fuel policy was associated, first, with the promotion of national unity through expanded interprovincial trade and, later, with the preservation of national independence through the achievement of self-sufficiency in fuel supply. Central to the realization of both these objectives was the problem of reducing the cost of transporting coal from the peripheral, producing provinces to the central, consuming ones, either by means of improved transportation systems or of freight rate subventions. Differences over fuel policy tended to resolve into controversies over transportation policy and, occasionally, the tariff.

Transportation policies also constituted the core of Canadian fuel policies in the age of oil and gas, and disputes over the most appropriate fuel transportation policies proved to be the most contentious issues in Canadian energy policy after World War II. (A large proportion of the major energy debates this country has had since World War II have involved the construction of pipelines.) While the objectives of national unity and national self-sufficiency in fuels continued to enter these debates to some degree, a more prominent issue for some time was the importance of preserving exclusive Canadian jurisdiction over transmission systems by insisting that they be constructed entirely within Canadian territory. "All-Canadian" routes were advocated to prevent American interference with deliveries to Canadian markets through U.S. territory and to avoid the development of an excessive rate of exports to American markets. However, Canadians often disagreed over the importance of these claimed benefits of exclusive jurisdiction. Interestingly enough, one of the reasons why the question of pipeline routes was so contentious was that the objective of maintaining exclusive Canadian jurisdiction over fuel transportation, desirable as it might have been in its own right, was often in conflict with the equally desirable objective of reducing the cost of transmitting Canadian fuels to Canadian markets, thereby improving the competitive position of Canadian fuels in Canadian markets.

Pipelines routed in part through the United States have frequently been said to have cost advantages over pipelines routed entirely within

Canada, whether they have been designed to carry Alberta fuels to central Canada or to the West Coast. When to this is added the fact that pipelines routed through the U.S. were likely to market a portion of their throughput in the United States, it is not surprising that numerous debates in Canada over the construction of pipelines pitted nationalists, who felt that Canadian markets should be served entirely with Canadian fuels and delivered by means of pipelines built exclusively within Canada, against "free marketers," who felt that Canadian markets should be served by Canadian sources only if necessary and then only by means of pipelines built along the cheapest possible routes. The former would maximize interprovincial trade at the expense of international trade, while the latter would do the reverse. The decisions actually taken and the transmission systems actually built resembled neither the nationalist nor the free-market, continentalist extreme. Viewed as a network of oil and gas transportation systems, the Canadian pipeline projects approved between 1949 and 1961 form what could be called either a quasi-national or a semi-continental pattern of fuel transportation and distribution, a pattern bearing a distinct resemblance to that established in the earlier era of coal, where Canadian production was roughly equal to Canadian demand but where a substantial portion of that production was exported while a substantial share of the Canadian market continued to rely on foreign supplies.

Canadian Pipeline Policy: An Overview

By 1958, for better or worse, the foundations of Canadian oil and gas policy had been laid, leaving room for few significant changes in subsequent years at acceptable political and economic cost. During the fifties, the federal government gave its approval to four of the major transmission systems to be built prior to the mid-1970s: the Interprovincial and Transmountain oil pipelines to carry Alberta's oil to Ontario and Vancouver markets respectively, and the Trans-Canada and Westcoast pipelines, to carry Alberta and British Columbia natural gas to central Canada and Vancouver respectively. The physical layout of these transmission systems was to constrain narrowly the disposition of oil and gas found surplus to Alberta's needs in subsequent years—once in place, it makes good economic sense to keep pipelines filled and even to allow them to expand their rate of throughput. Furthermore, the policies adopted by the federal government toward the construction of pipelines and the conditions under which they were to operate constituted a set of fairly explicit principles that have been part of government thinking ever since. Contrary to the views expressed by some critics of the federal government who portray Canadian energy policy as if it were haphazard and lacking any form of guiding principle, Cana-

dian governments have maintained some consistency in their approach
to pipeline development and oil and gas marketing. The best example of
this is what may be called the "joint service" concept.

This concept can be summed up as a policy of building an "export
component" into major Canadian pipeline projects in order to lower the
unit cost of transmitting Canadian fuels to Canadian markets, a saving
which results from the economies of scale to be derived from increasing
the volume of oil or gas which is put through a transmission system. The
returns to scale for pipelines are significant because of their high ratio of
fixed to variable cost, and because of an immutable law in the geometry
of all cylinders, including pipelines: For every increase in the diameter
of the pipe there is a much higher increase in the total capacity of the
line. (A study done for the federal government at an early stage of the
development of Canada's pipeline systems had concluded that about 70
percent of the cost of building a pipeline is fixed.[1]) As a consequence of
these factors, the larger the total volume of deliveries through a pipeline
system, the lower the cost of transporting each barrel of oil or each
thousand cubic feet of gas becomes. When one adds to these facts the
consideration that, at the time it was proposed to build various pipelines
in Canada, the Canadian market was small compared to the optimum
carrying capacity of the pipelines of the day, it is not hard to see the
economic rationale for including a fairly large capacity for export service in Canadian pipeline projects.

Each of the four pipeline systems named above included a significant export component. As we shall see, the economies-of-scale rationale was explicitly applied to both the Westcoast and Transmountain
projects. Exports were also added to the original Trans-Canada pipeline
as a means to facilitate its financing. The importance of exports in
establishing the feasibility of the Interprovincial oil pipeline was a
disputed point at the time of its approval, but they certainly formed an
important part of the project in its latter stages. Moreover, twenty years
later, the Canadian Arctic Gas Pipeline was argued by industry and
government spokesmen to be in Canada's interest on the basis of the
scale economies that could be achieved by bringing Canadian gas to
Canadian markets in conjunction with a project bringing Alaskan gas to
American markets. Although this project was later rejected, its suc-
cessor, the Alaska Highway Pipeline, was also initially recommended to
Canadians, in part, on the same grounds.

The 1950s, then, were a decade of rapid expansion of the oil,
natural gas, and pipeline industries in Canada with lasting effect on
Canada's energy development. Oil production increased from about 30
million barrels in 1950 to about 190 million in 1960. Natural gas production increased from about 70 million cubic feet to over 500 million
cubic feet over the same period. Five transmission systems—the Inter-

provincial Oil Pipeline, the Westcoast gas pipeline, the Trans-Canada gas pipeline, and the smaller-scale Canadian-Montana gas pipeline—were all incorporated and given federal approval prior to 1960. Two other major pipeline projects were considered and rejected or delayed during the same period—the Alberta Natural Gas and the Alberta-to-Montreal oil pipeline proposals. Finally, a royal commission on energy was appointed, held its hearings, and reported at the end of the decade. While certainly one of these developments—the Trans-Canada pipeline debate—can hardly be said to have received insufficient attention from students of Canadian energy policy, the others deserve more attention than they have been given, if only for the reason that they touch dimensions of Canadian energy and pipeline policy that either did not arise in connection with Trans-Canada, or elicited different policy responses from those to Trans-Canada.

The Pipe Lines Act

The most tangible evidence that major developments in the transportation and marketing of Canadian oil and gas were soon to demand the attention of the federal government was the introduction in April 1949 of a bill (eventually the *Pipe Lines Act*, R.S.C. 1952, c. 211) providing for federal control of interprovincial and international oil and gas pipelines.[2] Under the terms of this act, companies proposing to transmit Canadian oil or natural gas to markets outside the producing province were required to be incorporated by an act of parliament, following which details of the project, such as the route to be followed, would be placed before the Board of Transport Commissioners for approval or rejection.

Considering the length and, at times, the acrimony of later debates over the incorporation of companies under its auspices, the debate on the Pipe Lines Act itself was rather innocuous. The bill to establish the act was introduced by the Minister of Transport, who pointed out that the bill reflected the view of his officers that "to give private companies power over interprovincial and international pipe lines, without some governing body, would be to create a position of chaos in a new and growing field." He went on to state that "the importance of the oil and gas industries to the economic welfare of Canada cannot be over-emphasized."[3]

The opposition parties in the House were generally in agreement with the principles of the bill. Howard Green, Conservative member for Vancouver South, welcomed the bill, but pointed out that the federal government was moving into an area of provincial jurisdiction, and warned that the bill "would not work unless the utmost cooperation with provincial authorities was undertaken."[4] Green identified several major principles which should guide the administration of the act. He argued that Canadian oil and gas should be used in Canada as much as possible; that the Dominion government, having entered this field, should now be responsible for giving assistance to the provinces for the conservation of natural resources; that future oil and gas development must bring down the cost to Canadian consumers; that oil and gas pipelines must serve all comers; and, finally, that the government must set up an efficient staff under the Board of Transport Commissioners to administer the act in accordance with the expertise of industry oil men.[5]

Arthur Smith, the Conservative member from Calgary, questioned the quantity of gas available for export to other provinces or the United States, reflecting a concern over the security of supply for Alberta which was widely shared by his fellow Albertans at the time:

> We in Alberta agree that gas should be used elsewhere but only to the point where we ourselves are safe in the use of that which Providence has thrust under our feet.[6]

Smith added that he was "not sure where dominion jurisdiction leaves off and provincial begins," but declined to press the point, favouring instead "some central authority to deal with the matter since it is interprovincial in nature."

The position of the Conservative party with respect to the Pipe Lines Act reveals for the first time a number of factors that were to bear significantly upon later pipeline bills. Howard Green stated that the Conservative position on oil and gas development in Canada was based upon the principle that Canadian reserves should be used primarily to meet domestic requirements, while leaving open the option of exporting surplus quantities of oil and gas. He saw "no reason why eventually gas should not be piped down the Pacific coast."[7] He thought the probability of a surplus to be high and that any such surplus "could be sold in the United States." Although Green in this way clearly did not rule out exports, he did see them as a policy option only if the requirements of Canada were met first.

As we shall see shortly, the passing of the Pipe Lines Act raised the curtain on a drama involving numerous competing pipeline proposals—several of which were already in the wings and several others that were rehearsing their parts—and a series of lengthy and contentious parliamentary debates on their respective merits. Nearly all of these debates directly or indirectly involved disagreements over the extent to which Canadian oil and gas should (or could economically) be devoted to the meeting of Canada's own requirements, and the extent to which exports to the United States might interfere with this. To place this debate in the proper context, it is worth noting that the Pipe Lines Act was not the first occasion upon which the idea of taking Alberta oil and gas to central Canada by pipeline had come up. In 1947, for instance, an MP from southern Alberta called upon the federal government to see that pipelines were built to carry the natural gas from the Turner Valley field to central Canadian markets.[8] (Much of this gas was being "flared off" or "torched"—that is, indiscriminately burned in the field—in huge quantities, to the amazement of residents and tourists.) This suggestion, incidentally, was made in the course of a debate on the appropriate Canadian response to yet another threat to the delivery of American coal to Canadian markets in the form of either a possible American coal embargo against exports to Canada or a Great Lakes shipping strike. As for oil, an Alberta royal commission had found in 1940 that a pipeline from Alberta to the Great Lakes, aimed to serve refineries in Ontario, would be an "economically sound proposition" once oil reserves in Alberta were sufficient to assure a minimum daily throughput of 60,000 barrels a day for a period of twelve to fourteen years, or about 300,000,000 barrels in all.[9] Moreover, by the time the Pipe Lines Act was in place, the vulnerability of central Canadian markets to shortfalls in or suspensions of imported fuel supplies had been underlined by a combination of disruptions of coal supply, as we have just seen, and shortages of fuel oil.[10] Indeed, the year before the Pipe Lines Act was introduced, the member for Calgary East had referred to a possible American embargo of oil exports to Canada (a prospect that C. D. Howe had earlier conceded to be both highly possible and potentially a "calamity of the first order") and had called upon the government to provide tax incentives of various kinds to encourage exploration for oil in Alberta in order to "make us sure of our own supply by developing our own resources."[11]

As the sense of anxiety over Canada's fuel situation grew once again, so did confidence in Alberta's capacity to serve the needs of Canadians beyond the province's borders. Difficult problems remained, however, and they essentially paralleled Canada's coal problem in the 1920s; Ontario needed oil and gas; Alberta had oil and gas: again, it was "purely" a matter of pipelines. However, doubts about the "purity" of pipeline politics were to arise very quickly on the heels of the Pipe Lines Act.

Canadian Pipelines: 1949 Proposals

When Canada's pipeline drama opened with the passage of the Pipe Lines Act, three principal actors were already on stage, along with some minor ones. Interprovincial Pipe Line Company, Westcoast Transmission Company, and Western Pipe Lines (later to become part of Trans-Canada) were all incorporated within weeks of this action.[12] By the end of the year, Parliament was also considering the incorporation of Alberta Natural Gas Company and Prairie Transmission Lines Limited. Before describing the debates and decisions which these individual projects occasioned, it may be helpful to indicate what, essentially, each of these companies proposed to do and how decisions in one case were bound to affect the fate of the others.

First, and in this context simplest of all, was the Interprovincial oil project, which was intended to transport oil from Alberta to Ontario by means of a pipeline to the Great Lakes, whence the oil would be loaded onto lake tankers for shipment to refineries at Sarnia and other points in Ontario. The project had no rivals seeking to incorporate for the same purpose, and the controversy that was to develop around this proposal had to do almost exclusively with the routing of the pipeline and the location of its terminus on the Great Lakes: Fort William, Ontario, or Superior, Wisconsin?

The remaining four projects, all gas transmission pipelines, were at this early stage rivals in the fundamental sense that they each sought to be the vehicle through which Alberta gas would serve markets outside the province. To make matters more difficult, three of the companies proposed to serve, at least in part, the same market: Vancouver and the states of Washington and Oregon. Only Western Pipe Lines (later Trans-Canada) was designed to take Alberta gas eastward to Winnipeg and the midwestern states in the U.S. Thus, during the early period in which the amount of gas in Alberta surplus to the province's own future requirements did not promise to be very large, the project that was to evolve into the Trans-Canada pipeline had to compete with the other projects oriented toward West Coast markets; and for a time it looked as though a choice would need to be made between serving markets either to the west or to the east of Alberta. Beyond that, approval of any one of the western-oriented pipelines involved a choice between an all-Canadian route and routes running partly through the United States. Westcoast's original proposal amounted to an all-Canadian route from gas fields in Alberta and northeastern British Columbia to Vancouver, with an extension for export service into Washington and Oregon, whereas Prairie Transmission and Alberta Natural Gas both proposed lines running south of the international boundary from southern Alberta to Seattle, with spurs running north to Canadian towns along the way and, of course, a final spur running north to Vancouver. While the issue of all-Canadian versus American routes was itself contentious, it took on added intensity because of the objection raised in certain quarters that routes running south of the border entailed the possibility, if not the certainty, that service to Canadian markets was secondary to export service, rather than the other way around.

In reviewing the considerations that surfaced in Parliament on the bills to incorporate these companies, the discussion will focus, first, individually on the oil pipeline proposed for service to Ontario, and second, collectively on the several natural gas pipeline projects described. While the final decisions concerning some of these proposals were not made until much later, it is useful to review how they were conceived at the time of their incorporation and where they stood at the point when, in 1951–52, a second set of proposed pipelines emerged following the outbreak of the Korean War.

Interprovincial Pipeline
Interprovincial was incorporated on the basis of a plan that called for a crude-oil pipeline to be built from Edmonton to Regina and points east for the purpose of securing a market for newly discovered reserves of Alberta crude oil. By June 1949 the pipeline was completed to Regina, and by early autumn of that year the line stretched to Gretna, Manitoba, a point south of Winnipeg on the Manitoba–South Dakota border.[13] From Gretna, the pipeline crossed United States territory to its lakehead terminus at Superior, Wisconsin. Here the oil was shipped by tanker through the Great Lakes to refiners in Sarnia, Ontario. Only in 1953, when demand had reached a sufficient level, was the pipeline extended from Superior across the Straits of Mackinac through the Michigan peninsula to Sarnia.[14]

The location of the terminus in Superior occasioned considerable debate in the Commons over the desirability of constructing a system designed to carry Canadian crude oil to Canadian markets through American territory. For example, Howard Green questioned "whether the terminus of the main oil pipeline should be in the United States or the lakehead in Canada" and whether the government was wise to select an American route for the trans-shipment of Canadian oil to Canadian ports.[15] Moreover, Green produced figures to support his contention that the government had made allowance for oil exports in approving the American route, claiming the pipeline's capacity of 70,000 barrels per day was to be divided between a 35,000 barrels per day tanker shipment in bond to Sarnia and an equal amount to be refined at Superior. An export permit in the above amount from the minister of trade and commerce, dated September 13, 1949, was cited in support of Green's argument.[16]

Despite the contention of Imperial Oil that a $10 million saving was possible by placing the terminus at Superior, instead of Fort William, Green maintained that the "main pipeline should be laid in Canada" and "Canadian oil should be shipped from one Canadian port to another."[17] If it was not terminated at a Canadian port, Green pointed out, the pipeline would contribute nothing to the development of northern

Canadian communities. Finally, Green presented his most pressing contention that the Canadian oil shipped via U.S. ports "might be used to replace U.S. oil" as American supplies became depleted, thus leading to an increase in the volume of exports to the U.S. He noted in a parting shot that this would become a reality "if oil companies form policy rather than Parliament."

Green's objections fell on deaf government ears. C. D. Howe, Liberal minister of trade and commerce, defended the government's intention to terminate the pipeline at Superior, arguing simply that there was "a definite saving in the marketing of oil by that route."[18] Building to Superior would result in reduced construction costs and provide for oil exports to the United States. Similarly, the eventual extension of the line south of Lake Superior would keep oil transportation costs low compared with any possible all-Canadian pipeline, and would help reduce the price of oil in Ontario. During the course of this debate, Howe also made several statements indicating the government's general attitude towards oil exports:

> In my opinion international commodities such as oil that move freely from country to country and continent to continent, should not be confined by geography. The sensible way to market international commodities such as petroleum is to move them to markets nearest the source of supply. If this House of Commons should say to Alberta "we are sorry but we need that petroleum in Canada and you cannot market any in the United States until you serve all Canada," not only would the oil economy of Alberta be wrecked beyond repair, but the good relations that have existed between the United States and Canada with respect to the supply and distribution of petroleum would also be wrecked.

Howe further alluded to the economic advantages of dividing the Canadian oil market into two sectors, a scheme that was later embraced by the Royal Commission on Energy in 1959 and adopted by the Diefenbaker government in 1961:

> It is in the interests of users of petroleum to get that commodity from the cheapest source. For eastern Canada and perhaps in the Montreal area the cheapest source is still the Caribbean or the gulf ports of the United States. The best market for Alberta oil is that which could be made available nearest to the source of supply, and part of that market eventually may be in the United States.

In direct contrast to Howe's view was the connection that the opposition drew between the routing of the line and their desire to avoid export commitments before it was certain that Alberta would develop oil reserves in excess of Canadian requirements. According to George Drew, for example, our first obligation is to supply our own centres by means of a pipeline. Then let us make every barrel of oil which is surplus above our requirements available to our neighbours.[19]

While arguing for an all-Canadian route and Lakehead terminus, Drew acknowledged the past generosity of the United States in supplying Canada with oil, but enjoined Parliament to use common sense in determining the location of Canada's first major oil pipeline. Drew reasoned Canadians "could acclaim their own common sense in making sure our first oil line goes to a Canadian outlet, where ships from a Canadian port could carry that oil to Sarnia." In addition, those opposing the American terminus stressed the secondary benefits to Canada of locating the pipeline depot at Fort William, such as the eventual construction of an oil refinery there or a later extension of the line through northern Ontario to eastern markets.

It is worth noting that objections to both the American terminus and the concomitant prospect of oil exports to the United States were raised before the Board of Transport Commissioners during its brief consideration of the Interprovincial application. The case for the Fort William terminus and against the U.S. route was put before the commissioners in several telegrams from officials and private citizens from Lakehead communities and by the mayor himself.[20] Stacked against these pleas for an all-Canadian route were several other considerations recommended to the commissioners by (primarily) representatives from Imperial Oil. One of the points raised was simply that the Canadian route would kill the project, which, it was argued, could not operate efficiently without the exports facilitated by the extension of the line into the United States. There was the further claim that the U.S. route was shorter, crossed easier terrain, and was consequently cheaper. Moreover, C. D. Howe and the Department of Trade and Commerce endorsed the project in this form.[21] Whether primarily for this reason, or in recognition of the considerations already discussed, the Commission approved the line as originally proposed at the end of the second day of hearings.

Natural Gas Pipelines

Three gas transmission companies sought federal incorporation in order to transport gas from Alberta to the Pacific Northwest states of Washington and Oregon and to Vancouver in British Columbia. Of these, Alberta Natural Gas Company and Prairie Transmission Lines were subjected to severe criticism—even obstruction—by opposition members, primarily because of their apparent intention to run their main lines south of the international border. For example, Douglas Harkness, a Calgary Conservative, led an attack against the Alberta Natural Gas proposal as one that was "prejudicial to the interest of

British Columbia, Alberta, Calgary and Canada."[22] He condemned the American routing of the proposed line, pointing to the possible escalation in the price of gas in both Alberta and Vancouver that could result from an increase in exports. Moreover, to leave Vancouver at the end of a pipeline routed through the United States would cast doubts on the security of supply for the Vancouver market. As Harkness put it, there was "not much question that once the gas was in the U.S., it would go to U.S. cities."

Supporters of Alberta Natural Gas tended to respond to this with the argument that American routes and exports to the United States worked to the advantage of the Canadian consumers to be served by such projects as Alberta Natural Gas. For example, J. S. MacDougall, a Vancouver Liberal, was not opposed to the pipeline, but was instead "only interested in seeing natural gas available in Vancouver through the shortest and cheapest route, even if it crossed through the United States." As for the higher export potential of American routes, another B.C. Liberal pointed out that

> gas cannot be profitably sent into that part of the country [B.C.] with its small population without the surplus being exported to the U.S. It does not matter who builds that pipeline in B.C., the company will have to send the surplus to the United States.[23]

Eventually, concern over the effect of the proposed project on natural gas prices and the growing feeling that Canadian needs should be served first prompted a CCF member to propose the following amendment:

> That Bill 66 (Alberta Natural Gas Co.) should not be read a second time, but deferred until the House has been assured that the route of any pipeline . . . will be laid out so as to serve Canadian requirements before any such pipeline leaves Canadian soil.[24]

Despite the support of the Conservatives, some western Liberals, and Social Credit members, this motion was defeated. Nevertheless, an opposition filibuster caused the Alberta Natural Gas Bill to die at the end of the 1949 session. February 1950 saw it reintroduced in substantially the same form, but the opposition continued to block its passage, re-emphasizing their previous arguments. Despite these prolonged efforts on the part of the opposition parties, the bill to incorporate Alberta Natural Gas Company passed Parliament by a wide margin. However, it had taken the bill some eight months to navigate the storm of Commons debate.

Similar parliamentary reactions were occasioned by the bill to incorporate Prairie Transmission. This is not surprising, since the basic design of the project—despite ambiguities surrounding the actual route that was being proposed—seemed almost identical to that of Alberta Natural Gas: The company proposed to carry gas from Alberta to the West Coast, serving points in southern B.C. along the way, with Portland and Vancouver as dual termini. Again, the bill was criticized because it did not guarantee that the pipeline would not be built south of the border or that exports would only be made after Canadian needs had been completely met. Thus, some of the speeches made during the discussion of this project amounted to repetitions of arguments made against Alberta Natural Gas.[25]

Several points made during the Prairie Transmission debate deserve attention, though, for they clarify what was at stake in the choice between the competing natural gas proposals of the day. One significant instance was the growing realization that there was little possibility of more than one pipeline being built to serve the West Coast. For this reason, some argued that the most desirable project was that planned by Westcoast Transmission, since it was the only one certain to be built across Canadian territory and the only one designed to tap the resources of the Peace River district of Alberta and British Columbia.[26] Not all members saw this feature as an advantage, for it implied that once all the various pipeline proposals had been decided upon, there would be only a single line transmitting gas east from southern Alberta and another single line transmitting gas west from northwestern Alberta and northeastern British Columbia. That is to say, there would be two companies, each with exclusive buying power in one of the two major gas-producing areas.[27] Thus, a member from southern Alberta, along with C. D. Howe himself, supported Prairie Transmission for the very reason that it would prevent a monopoly from developing in the business of transmitting southern Alberta gas.[28]

Another point in connection with both Alberta Natural and Prairie Transmission deserves mention, especially in relation to the Westcoast and Western Pipe Line proposals. While the notion of an American route for the westward movement of Alberta gas was rejected by many MPs, the resistance to gas exports as such was much weaker, provided Canadian needs were taken care of first. There was further the explicit recognition by some that exports in some volume were necessary if a line was to be built to Vancouver at all. For example, Art Smith quoted testimony before a Senate committee to the effect that "it is necessary to include the United States market because the limited market available in Canada would not, in itself, support the cost of a pipeline from any of the known Canadian natural gas fields to Vancouver."[29] It is probably for this reason that, despite the controversy that was to surround the Westcoast project at a later stage with respect to its methods of financing and the relationship between its export and Canadian pricing provisions, the Westcoast incorporation received relatively little attention from members of Parliament. Also, it did not raise the concern which, as we have seen, caused the greatest consternation among members of Parliament during this period of pipeline development, namely, its route. At least one member declared explicitly that Westcoast had been passed in the House "without any opposition whatever" because it was designed in a manner which assured priority of service to Canadian communities.[30]

Similarly, despite the controversy that was to be generated by the final form of the Western Pipe Line proposal (Western may be seen as the embryo of the export component of the ultimate Trans-Canada pipeline project) it is somewhat surprising that the bill to incorporate this company passed second reading in a matter of minutes.[31] This lack of opposition, together with the fact that the proposal was designed in part to export gas to the United States after serving markets on the Canadian prairies by means of an all-Canadian route, shows that criticism of pipeline proposals was almost exclusively restricted to the issue of the routes they were to follow. In sum, the evidence reviewed here would seem to confirm an assessment of the role of the opposition parties during this initial phase of the politics of pipelines in Canada:

> The main reason for the obstruction of the pipeline bills was the determination of certain members to ensure that any gas pipelines authorized by Parliament would be built entirely in Canadian territory. For them, the only justification for gas entering the United States would be to make the project economically feasible. However, their first priority was to see that Canadian gas was used to further Canadian development.[32]

Canadian Pipelines, 1950–1953

At the start of the 1950s, it was fairly clear that Interprovincial Oil Pipe Line would proceed much as originally planned, but that all of the natural gas pipelines would have to await word from the province of Alberta that the gas they proposed to transmit from the province to Canadian and export markets was, in fact, surplus to the requirements of the province. The Board of Transport Commissioners had ruled in September 1949 that a company required the Alberta government's permission to export gas from Alberta before the BTC could hear its application for permission to build a pipeline.[33] Meanwhile, the Alberta government was proceeding cautiously on the issue of exporting gas from the province, and the BTC received a statement from Premier Manning saying that a surplus of gas beyond the present and future domestic and industrial requirements of the province did not exist at that time "and is not likely to exist for some time to come."[34] This judgment on the part of the province was strongly influenced by the report of a provincial commission established in November 1948 to inquire into the proven and estimated reserves of natural gas in the province, and to investigate the present and estimated future consumption in the province.[35]

The report of the Dinning Commission had been received by the Alberta cabinet in March 1949. The commission reported that it had been impressed with the consistency with which parties to the hearings expressed the view that the people of the province should have prior claim on provincial gas supplies and that Canadian users should take priority over foreign users when and if a surplus developed in the province, a principle which the commission supported.[36] Presumably influenced by this observation and recommendation, the province proceeded in 1949 to enact the Gas Resources Preservation Act and, later in the same year, to establish a Petroleum and Natural Gas Conservation Board to regulate the removal of gas from the province. As a consequence, the fate of the proposed gas pipelines whose federal incorporation was just reviewed lay initially in the hands of the new provincial authority. As we shall see shortly, its work was complicated by the appearance of several new pipeline proposals and by the advent of the Korean War.[37] The second act of the play before us opens with the appearance of four major new actors: Transmountain Oil Pipeline, designed to carry Alberta oil to the Pacific coast of Canada and the United States; Border Pipeline Corporation, which wanted to do the same thing by an American route; Canadian-Montana Natural Gas, designed to supply the mountain states of the U.S. with Alberta natural gas; and Trans-Canada Pipe Lines, Ltd., which submitted a proposal for an all-Canadian natural gas pipeline to carry Alberta gas to Toronto, Montreal, and Ottawa. Once federal and provincial authorities had decided which of these projects and those previously described was to survive and which was to fail—a process that was essentially completed by 1953—the foundations of the oil and gas transmission network in Canada would be set.

Transmountain and Border pipelines
The Transmountain and Border pipelines are considered together because they planned to do basically the same thing—convey Alberta oil to west coast markets in Canada and the United States—by two different routes: Transmountain by an all-Canadian route through the Yellowhead Pass to Vancouver and Seattle, and Border by a route just south of the international boundary in a manner similar to the Alberta Natural Gas project. The Border Pipeline debate will not be discussed here in any great detail, since it produced few new arguments on the value of all-Canadian routes; but it deserves passing mention because it is a good indication of where pipeline matters stood in Parliament in 1951. In

brief, the opposition had developed a strategy to "talk out" any bill to incorporate a company that did not include the provision that "the main pipe line or lines for the transmission and transportation of gas and oil shall be located entirely within Canada."[38] Indeed, in a brief debate on another pipeline bill, Border was singled out as the "one pipe line bill before the house at present which does not meet that requirement."[39] The opposition was at pains to point out that bills that did meet this requirement owed their rapid progress through the house to the cooperation of Howard Green, to which an apparently disgruntled government member responded, "You mean that he did not block them."[40]

Transmountain therefore not only had the advantage of proposing from the start an all-Canadian route from Alberta to Vancouver, for which its sponsor was commended,[41] but was also provided an additional boost by the urgency of fuel shortages on the west coasts of Canada and the United States brought about by the Korean War.[42] The project's sponsor, Arthur Laing of Vancouver, argued for the pipeline in part as a contingency against the "eventuality where, through the impairment of Middle East oil facilities, we could have Venezuela oil and California oil immediately drawn to Europe and could be left without oil."[43] The chief commissioner of the BTC declared in April 1951 that the board, which would have preferred to adjourn until September, could not do so "in view of the Defense program both in this country and in the United States, and the near emergency which exists . . ."[44] This decision presumably expedited the progress of the Transmountain project, since a potential competitor (which had requested a postponement in proceedings on the project because it had been unable to obtain technical data necessary to make its own application) was told by the commission that the hearing on Transmountain would take place before the board would consider a postponement.[45]

Canadian-Montana

The bill to incorporate Canadian-Montana Pipe Line Company, introduced in the House on March 13, 1951, proposed the construction of a natural gas pipeline from the southeastern portion of Alberta to Montana. Gas purchased by Alberta and Southern Pipeline Company would be transmitted through existing pipelines to a point in southwest Alberta, where it would enter the Canadian-Montana pipeline for delivery to the Montana Power Corporation at a point near Cardston, Alberta. The main incorporators of the project, McColl-Frontenac Oil Company and the Montana Power Company, sought permission to construct the line primarily in order to serve the Anaconda Copper smelter in north central Montana.[46]

Incorporation of this project, as with Transmountain, was spurred on by the Korean War. This export of natural gas from Alberta, the first such export to be approved, was made at the request of the American military and the Canadian Department of Defence to meet emergency gas requirements of the metal industries in Montana.[47] This necessitated a constant supply of natural gas to keep Anaconda in operation, as the costs in time and money for Anaconda to convert to alternate fuel supplies would have resulted in a decrease in the production of necessary war materials. To that end, the U.S. government asked Canadian authorities to intercede with the Alberta government and grant the necessary export permits.

Colin Bennet, Liberal parliamentary sponsor of the bill, argued that the taking of gas from newly discovered fields some thirty miles from the Alberta-Montana border "will not particularly affect Alberta's supply."[48] He argued that the gas fields were already under the control of the pipeline sponsors, and that the application differed from others as the amount slated for export was smaller than other proposals. Douglas Harkness protested this incorporation in terms similar to those presented in previous cases. He attacked the bill by arguing that the incorporators had tried to show in committee that the only market for this gas lay in Montana, whereas gas from the same field could in fact be used for the anticipated Trans-Canada pipeline project to move gas east to Toronto and Montreal and for local domestic consumption in Calgary and adjacent communities. This position was similar to the stand taken by the Alberta government and the other western MPs.

The Canadian-Montana debate was similar to the Transmountain debate, in that it became intermingled with the rhetoric and emotionalism of the continental defence arrangements between Canada and the United States. Edward Applewhaite, a B.C. Liberal, noted that "gas is vital to the defence efforts of the United States," and that "this matter is in the interests of [Canada]," and urged members "to waive their rights [of debate] and allow the bill through."[49] Howard Green led the parliamentary opposition to Canadian-Montana. He pointed to a second application by the sponsors "for the use of Alberta gas for civilian purposes in the state of Montana," noting the potential danger that "Canada may go short." Green flatly proclaimed that the "Montana affiliate of Canadian-Montana has the intention to import Canadian gas to other U.S. centres," using the same gas field Trans-Canada hoped to rely on for transmission to central Canada. He voiced the hope that movements of this kind would be watched, as "there may be further attempts by other companies to pipe gas to Minneapolis and St. Paul." Opposition also came from the government side when John Sinnot, a Manitoba Liberal, bitterly complained that the "sponsors of this bill would export these natural products from this country" and, as a member of Parliament, he "would not stand for it."[50] He echoed opposition sentiment that "natural resources should be developed for the people of Canada, before they are exported anywhere else."

In the end, the federal and provincial governments acceded to the application of Canadian-Montana by granting a special export permit. The Alberta government enacted a special law allowing the removal of 23.8 billion cubic feet of natural gas per year at daily rates up to 40 million cubic feet.[51] The federal permit to export, issued in 1951, covered gas deliveries to Canadian-Montana for a five-year period.

Trans-Canada

Considering the storm the Trans-Canada project was one day to generate in the House of Commons, parliamentary response to its original bill of incorporation would not even pass for a distant roll of thunder. Support for the bill was almost unanimous in light of the fact that it proposed an all-Canadian route for the transmission of gas from Alberta to central Canada. The single exception to this unanimity was a member from Cape Breton, who feared that the proposal to market Alberta gas as far east as Montreal and Quebec City would threaten the sale of Nova Scotia coal in the same market.[52] Besides providing for an all-Canadian route, Trans-Canada voluntarily accepted the idea of embodying a commitment to this effect in its bill of incorporation, the first such company to do so. The proposal made no reference to the prospect of exports to the United States, a point which was favourably received as well, in that it met the priority given to Canadian over export markets by Alberta's Dinning Commission.[53] Several references were made to the opportunity provided by the project to reduce Canada's dependence on American supplies of coal and oil. And Fort William would finally get a pipeline.

The story of how, after this auspicious beginning, Trans-Canada became the instrument of Diefenbaker's rout of the Liberals in 1957–58 cannot delay us here.[54] In the context of this chapter, however, it should be pointed out that before Trans-Canada received permission to take gas out of Alberta, it amalgamated with Western Pipelines in March 1954 "as the result of some persuasion from the Alberta Government and the Government of Canada."[55] As a result of this combination of the initial all-Canadian project with Western's project designed to export to the U.S. at Emerson, Trans-Canada was deemed a more efficient proposal. But it still required public assistance to get off the ground, and this was where its real troubles began. So extensive was the public assistance involved—the federal and Ontario governments were to build the Ontario portion of the line north of the Great Lakes, and the federal government was to provide short-term financing for 90 percent of the cost of the Prairie section—that large numbers of Canadians objected to leaving the control of the pipeline in private hands (and American private hands, at that). There were also some concerns about the export component of the proposal. But there seems to be agreement among most observers that the procedures employed by the government to force the legislation through the House had more to do with its later political difficulties than the content of the policy pursued.[56]

The Outcome

Specific details of the disposition of the various proposals put forward by the companies whose incorporation we have reviewed would require a close reading of the proceedings and reports of the federal BTC and the provincial conservation board. But the effect of the decisions taken by those agencies is not hard to summarize. With respect to oil, Canada was to acquire one pipeline to central Canada and one to the west coast. With respect to natural gas, precisely the same was true. All four pipelines were designed to serve export as well as domestic markets and, indeed, depended on exports to be viable. (As a small distortion of this neat arrangement, it is necessary to note the existence of Canadian-Montana, an exclusively export-oriented venture aimed at a small regional market in the U.S.) The surviving actors were Interprovincial, Transmountain, Trans-Canada (Western), and Westcoast. Lying dead or wounded on stage were Prairie Gas, Border Pipe Lines, and Alberta Natural Gas, along with some bit players and extras whose stories add nothing of significance.

Border would appear to have lost to Transmountain on the grounds that it seemed likely to follow a route through the United States and, in any case, was not in stable shape at a time when the Korean War made a pipeline to the west coast a matter of some urgency. Prairie Gas and Alberta Natural Gas fell to the 1952 decision of the Alberta Oil and Natural Gas Conservation Board that the gas in the northwest part of the province, but not that in the southern part, was surplus to the requirements of the province, and that Westcoast should be permitted to deliver this surplus to west coast markets.[57] A permit to this effect was issued to Westcoast by the Government of Alberta (O.C. 882/52) on June 16, 1952. Since Westcoast was thus on its way to becoming the means by which the west coast was to be served with natural gas, and since the market would support only one such line, both Prairie and Alberta Natural Gas were deprived of their *raison d'etre*.

Similarly, Trans-Canada was to become the sole means by which central Canada was to be served with natural gas, despite some last-minute efforts on the part of both Prairie Gas and Alberta Natural Gas to promote new proposals to supply central Canadian requirements. Alberta Natural Gas proposed to do so by means of an all-Canadian pipeline.[58] Prairie proposed to do so by arranging a ''gas swap,'' whereby it would export gas to the west coast of the United States, and in return a Texas company would make provision to supply eastern Canada.[59] This proposal for an international exchange was strongly opposed by other applicants on the grounds that it was probably impossible given that it would require revisions in statutes of the United States prohibiting uninterruptable exports of natural gas from that country.

This reference to the likely impossibility of a gas-swap arrangement, whether under the auspices of Prairie Gas or any other company, is interesting, for a few years later C. D. Howe was to argue that an all-Canadian gas pipeline to the central provinces, though less desirable on most counts than a continental swapping arrangement between Canada and the United States, was a virtual necessity given the fact that Canada could not count on long-term imports of gas from the U.S. It is true that Howe went to great lengths to prevent an American company from serving the central Canadian market with gas from the United States, after the Federal Power Commission had approved uninterruptable export sales to the Canadian market.[60] However, this apparent "protectionism" on Howe's part should be seen as primarily a tactical means of ensuring the successful completion of Trans-Canada, while the strategic importance of the all-Canadian project itself stemmed from the refusal of the American authorities to commit American gas to meeting future Canadian requirements. Even the U.S. export decision that caused Howe so much concern strongly disavowed any implication of a commitment to future U.S. gas exports to the same or to new Canadian markets.[61]

The Government Pipeline Statement of 1953

The previous section broadly describes the situation in Canada with respect to pipelines as it emerged during the period 1950–53. Two of the projects, Westcoast and Trans-Canada, would encounter some tough sledding on their way to completion later in the decade. In particular, each would be subject to investigation by the Royal Commission on Energy struck by John Diefenbaker almost immediately upon his election in 1957. Interprovincial, though, pushed its way without difficulty to Superior and, in a few years, on to Sarnia. Even so, its owners would become involved with the hearings of the Energy Commission, where they attempted to block a proposal for an all-Canadian oil pipeline from Alberta to Montreal. Only Transmountain and Canadian-Montana were to get on with their appointed business outside the realm of public controversy.

In short, there was some unfinished business in 1953. As if to sum matters up, the St. Laurent government made a major policy statement regarding energy. The policy enunciated in the House by C. D. Howe on March 13, 1953 was divided into two sections, one concerning gas, the other concerning oil. In the preamble to his statement, Howe emphasized the constitutional provision delegating resource ownership to the provinces, but he also stressed the power of the federal government over international and interprovincial trade.

With respect to petroleum, Howe stated that Canada's policy should be twofold: First, "to move petroleum from the source of production to refineries within economic distance in the cheapest possible way"; second, "to arrange for markets for that portion of Canadian output that cannot be economically used in Canadian refineries in the market that offers the highest return to the producer."[62] Howe further stated that this policy was consistent with previous government attitudes toward crude oil, with the Interprovincial pipeline project providing the best example. Similarly, the Transmountain pipeline to Vancouver, in Howe's words, "does not violate government policy, provided refineries at Vancouver will first obtain an adequate supply of petroleum from Alberta." This policy on crude oil was essentially a restatement of the 1949 energy speech given by Howe during the debate over the Lakehead terminus location of the Interprovincial pipeline. The need to move crude oil to market "in the cheapest possible way" included the necessity of achieving economies of scale through the provision of exports or through American pipeline routing or both. Similarly, the concern for markets makes it clear that, at this stage, the question of crude oil exports was a foregone conclusion. This is clearly evident in Howe's remarks. If it was cheaper to pipe Canadian oil through the U.S. and to export some along the way, then the economics of transportation dictated that it should be so. Howe seemed to close the door on a nationwide market for Canadian oil. Because the distance to the east of Canada involved such high transportation costs, the solution in Howe's mind was to seek the nearest available market. That market could only be the United States.

Government policy toward natural gas was somewhat different. Howe compared the trade in natural gas to that in electrical power, declaring that "the policy that guides the export of electricity since 1907 applies within reason to the export of natural gas." Thus, it was government policy to refuse permits for moving natural gas by a pipeline across international boundaries until the government was convinced "there can be no economic use present or future for that gas within Canada." Referring to the Westcoast transmission and Canadian-Montana pipeline projects, Howe remarked that these projects were the only exceptions to government policy. In the case of Canadian-Montana the decision was made to supply the Anaconda Copper refinery in Montana with gas to ensure the continued production of necessary war materials. In the case of Westcoast, export permission was given because of the gas shortages in the American Pacific Northwest and the insufficient demand in the Vancouver market, which made this area in the U.S. a logical market for Westcoast's northern gas fields. Despite a high degree of concern in Alberta and British Columbia for security of natural gas supply, it was found that "ample amounts were available to serve the Pacific coast, and guarantee a substantial export of gas to the northwestern states and in addition, to serve the present and future needs of B.C." Howe argued that, as with oil, "this is the only practical way of assuming early delivery of natural gas to the industrial areas of B.C."

The most significant point of government policy dealt with natural gas supplies to Ontario and Quebec. In Howe's words, the "only reliable supply of natural gas for the provinces of Ontario and Quebec must be from western Canada by means of an all Canadian pipeline." He explained further that it was "economically feasible to build a gas pipeline from southern Alberta to Toronto and Montreal," and "government policy will require that Canadian gas be used in Canada accordingly." With this statement, Howe appeared to have achieved a major political coup. Opposition parties welcomed the new energy statement with favourable comment. The only unfavourable response came from the CCF, which once again called for public ownership of gas transportation facilities. Otherwise, all parties were in substantial agreement.

It seems ironic, therefore, that C. D. Howe got himself and his government colleagues in so much trouble over the "Great Debate" on the Trans-Canada project in 1956, a project that departed furthest from the design that would have better suited his general continentalist orientation. It seems fair to say that Canada owes to Howe's determination the fact that Trans-Canada Pipe Lines exists at all. Trans-Canada, bringing as it eventually did an Alberta fuel source all the way to Montreal, thus serving both the major fuel markets in the country, constituted the closest thing to a nation-wide fuel market the country had seen or would see until Alberta oil arrived in Montreal in 1974. The official justification of this project was to guarantee central Canada reliable access to a premium fuel. The opposition had little quarrel with this. What seems to have been at issue was the necessity of the extent of public assistance Howe seemed determined to provide the (American) promoters of the venture and his use of closure to settle the question. There was also growing concern about the possible threat to future Canadian requirements represented by the export component of the project. However, following the final go-ahead for Trans-Canada in 1956, the principal importance of the project from the viewpoint of this study is the impetus it provided the new Progressive Conservative government to establish a royal commission on energy to inquire into a host of concerns generated by the pipeline politics of the 1950s. Westcoast had also raised concerns about the advisability of gas exports to the United

States, and the price at which these exports were made; and these matters, too, were to be examined in detail by the new royal commission.

Thus, attention must turn to the Royal Commission on Energy (the Borden Commission). This is partly because the commission considered and ultimately rejected the Alberta-to-Montreal oil pipeline, the last major pipeline decision taken during the first decade of policy making with respect to the disposition of Canadian oil and gas. The commission was also instrumental in establishing the federal government's approach to the issue which dominated the second decade of oil and gas policy, namely, the rate and prices at which oil and gas exports (especially gas) should be permitted. The Borden Commission not only recommended the creation of the National Energy Board, but established a procedural and conceptual legacy for the board with respect to the regulation of the oil and gas industry. Moreover, the board's subsequent exercise of its powers, partly in keeping with its legacy from the commission, eventually produced a situation in which yet another enormous pipeline project would be proposed in the early seventies, and would be justified as a means of reducing Canada's dependence on foreign supplies of fuel. There is no straight line to be drawn from the Borden Commission to the Alaska Highway project, but a line exists nevertheless.

Nationalism Versus Continentalism: The Borden Commission and the Marketing of Canadian Oil and Natural Gas, 1958–1960

DECISIONS made by federal authorities between the years 1949 and 1958 reveal that some basic policy attitudes and principles did govern the approach of the Liberal government and federal regulatory agencies to the questions of oil and gas exports and pipeline construction. One such principle—that oil and gas surplus to Canadian requirements ought to be exported to the United States—claimed the support of nearly all parties and members of Parliament. Such exports were deemed desirable on three counts: first, they stimulated the growth of the industry in Canada, thereby promoting regional development and, at the same time, assuring protection of Canada's long-term needs through a high level of exploration and development of Canada's potential fuel resources. Second, when exports were combined with the service of Canadian markets by means of the same pipeline facilities, they permitted the realization of scale economies that reduced the unit cost of transmission of Canadian oil and gas to Canadian markets and, in turn, promoted wider Canadian use of Canadian fuels. Third, pipelines serving Canadian markets with Canadian fuels should be built predominantly, if not exclusively, through Canadian territory. These principles emerged and were first applied during an intensely active period of pipeline construction, and were subsequently used to determine the acceptibility of different and often competing pipeline proposals. After 1958, primarily because four major transmission systems had already been approved, emphasis in national oil and gas policy shifted away from the basic design of transportation networks to the terms and conditions under which Canadian fuel resources were to be exported—the appropriate volume and price of Canadian exports. One contemporary observer summarized the problems facing the Canadian oil and gas industry and the federal government at the time:

> Canadian resources of petroleum are directly competitive with those of the United States. There are, therefore, no prospects of attaining a monopoly in this product; on the contrary, a serious problem exists of securing stable access to the United States market. On the other hand, exploitation of Canada's

resources of natural gas promises to contribute very substantially to the basic energy resources for Canadian manufacturing. There is, therefore, a problem of ensuring that the long-term energy requirements of the Canadian economy are adequately provided for, and that the industry does not become too dependent on United States outlets. Access to the American market is the primary problem for petroleum; restricting exports to the United States in the interest of the manufacturing complex of central Canada is the primary problem for natural gas. Public policy has differed considerably between the two products.[1]

From the late 1950s onward a controversy developed over the extent to which Canada's oil and gas (particularly gas) ought to be exported; it was perceived by increasing numbers of Canadians that, in the absence of effective government regulation, the Canadian petroleum industry might raise exports to a level that could either threaten the security of supply for Canadian consumers or restrict the number of Canadian users of Canadian fuels. Particularly because of the high level of American control in the Canadian oil and gas industry, there were fears that, left to themselves, gas producers and transmission companies might not show sufficient respect for the Canadian interest in preserving the country's supply of an increasingly important source of fuel that was not obtainable on a reliable basis anywhere but from western Canada. This conflicted with the desire of the industry to expand production around sales to American markets because these were generally larger and more lucrative than the new and remote central Canadian one. (It should be noted that this preference for the export market over the Canadian market was shared by Canadian firms.) Therefore, while it is true that consensus had already formed around the idea that quantities of oil and gas surplus to Canadian requirements should be exported, this consensus gradually broke down once it became clear that the significant issue was not whether surpluses should be exported, but how an "exportable surplus" should be defined. At the core of this controversy were some largely (but not entirely) technical issues—the size of Canada's established reserves, the size of Canada's present requirements, and the anticipated rate of increase of both reserves and requirements. But the real core of the controversy was made up of purely political questions, such as the length of time over which Canadian requirements ought to be estimated (that is, how much protection is enough?), the extent to which future Canadian requirements ought to be protected out of established reserves as opposed to future discoveries, and the extent to which exports ought to be relied upon as a means of stimulating exploration and development in the Canadian industry. The first agency to attempt to provide answers to such questions was the Royal Commission on Energy (whose chairman was Henry Borden,

C.M.B., Q.C., President of Brazilian Traction, Light and Power), and the proceedings of the Commission provided a forum for a full debate among the parties with conflicting interests in the matter of defining the size of the Canadian market for Canadian oil and gas.

The Appointment of the Borden Commission

The first substantial action undertaken by the newly elected Conservative government of John Diefenbaker, announced on the second day of the Twenty-Third Parliament, was to establish a royal commission to enquire into and make recommendations concerning a broad range of energy issues.[2] As we have seen, an extraordinarily vituperative parliamentary debate on Howe's Trans-Canada bill and an unusually powerful upsurge of dissatisfaction with the previous Liberal government had preceded Diefenbaker's striking of the Borden Commission. It is difficult to say to what extent this public mood reflected a strong desire for new departures in policies toward oil, natural gas, and pipelines, and to what extent the Canadian public had simply resented tactics and procedures employed by the Liberal government in the course of the pipeline debate. It is clear that the role of American firms, financial interests, and regulatory decisions did contribute to the government's political embarrassment over Trans-Canada.[3] In any case, the commission undertook its task of redefining Canada's oil and gas policies in the midst of a level of public concern whose intensity and scope would not be seen again for the next fifteen years.[4] It is also noteworthy that by this time the Liberals, Conservatives, and CCF all supported the principle of some comprehensive energy authority with powers sufficient to ensure that Canadian interests were not harmed by an industry in which foreigners played such a large part, although the parties differed on the nature of the powers required. The Borden Commission warrants close examination, for it marks a pivotal phase in the creation of the National Energy Board, the agency that was to stand at the centre of the energy controversies of the next two decades.

The idea of a national agency with the power to regulate Canada's energy trade did not originate with the members of the Royal Commission on Energy. As early as February 1955, Progressive Conservative members (then still in opposition) had called on the government to consider a national energy board, insisting that there was "a great need for an over-all energy policy" in order to do away with "dealing with the problem piecemeal."[5] By 1957, similar sentiments and support for a national energy board were voiced by numerous members of Parliament, owing in large measure to the fact that such an agency had figured in the recommendations of the Royal Commission on Canada's Economic

Prospects (the Gordon Commission), which released its findings in that year. A study prepared for the Gordon Commission had warned of the dangers to Canadian interests represented by the foreign ownership of companies engaged in the transmission and export of natural gas:

> The situation with regard to exports into the State of Montana illustrates perhaps in exaggerated fashion the lack of awareness in this country both of the broader Canadian interest in these matters and the intricacies of parent-subsidiary trade spanning the International Boundary. . . The Canadian subsidiary of the Montana Power Company has been permitted to charge its parent a price at the International Boundary comparable to that prevailing in other fields in Alberta. Canadian gas, priced in this way, is being delivered at the Anaconda Smelter in Butte, Montana, at a cost to that company between one-half and one-third of that which it would otherwise have had to pay for energy in the form of coal or residual oil. Had the forces of supply and demand been allowed free play, the return at the producing level would therefore have been higher.[6]

The same study went on to point out that American importing companies owned stock in the two other exporting companies in Canada, Trans-Canada and Westcoast Transmission. "These customer companies, because of their corporate ties, have exerted (and may well continue to exert) considerable influence upon the pricing and other policies followed by their subsidiary and other supplier companies in Canada."[7] The concern of the Canadian public over this question of the export-pricing policies of foreign-controlled companies had been sharpened, particularly by the price—22 cents per thousand cubic feet, with no provision for escalation over a twenty-year period—at which Westcoast was exporting gas to the American Pacific Northwest.[8]

In its final report, the Gordon Commission recommended the establishment of a national energy authority with advisory and regulatory powers. It also warned of the extensive foreign ownership and control of the petroleum industry, pointing out that in the board rooms of international corporations "the horizons are wider and the pressures different—although not necessarily all less parochial—than they would be in Canada."[9] Nevertheless, the Gordon Commission foresaw a period of time in which Canada's exports of oil and gas would greatly increase.[10] John Diefenbaker, still leader of the official opposition, had endorsed the Gordon Commission recommendation for a national energy board for similar reasons. During a major policy speech on the need for a national development policy, he criticized "not American investment in Canada but the degree to which the investment in Canada by foreign corporations was uncontrolled for the benefit of Canada," and advocated "a plan; not a planned economy but a national policy;

not a policy of nationalism but one whereby Canada in the days ahead will remain an independent Canada and will not inexorably drift into economic continentalism."[11]

Given this sort of background, it might have been expected that Diefenbaker's new government would have incorporated into the commission's terms of reference some hint of a new direction in energy policy, the practical implications of which it would be the job of the commission to explore. However, it is evident from the vague generalities in which the terms of reference of the commission were set out that the government had not in fact decided the substance of a new national energy policy. The government appears to have entrusted the commission with the task of deliberating on and formulating the basic direction of Canadian energy policy itself.[12]

The government did favour the institution of a national energy board to administer whatever might be the eventual provisions of the new energy policy. As to the substance of the policy so administered, the attention of the commission was directed first toward the adequate protection of Canadian energy requirements, and second to something called "the most effective use" of Canadian resources "in the public interest," phrases which were left undefined in the commission's terms of reference. The terms of reference might just as well have been left blank, for "effective use" could be interpreted to mean, among many other things, a priority given to homeowner consumption, industrial consumption, thermal-electric production, short or long-term cash return, a guaranteed long-term surplus to ensure low prices to consumers, a high rate of exports to ensure higher prices to producers and continuous resource development, or the development of Canadian industries around an assured supply of low-cost premium fuel.

This list sums up many of the interpretations given to the phrase, "the most effective use in the public interest," by various groups interested in the commission's findings. There obviously must be some means of discriminating among the variety of policies that could realistically be claimed to protect Canadian requirements and to promote the most effective use of Canadian resources. Presumably, the phrase, "in the public interest," was added to the wording of the government's statement in order to qualify the meaning of effective use, but in fact it could not do so, for the notion of the public interest is as ambiguous as the notion of effective use. The job of the commission was not simply to recommend "the policies which will best serve the national interest" but was in fact to *define* the national interest. Because the specific content of the "national interest" in energy matters was left to the commissioners to define, the various formulations presented to it by interested parties during the commission's hearings deserve close attention. The key question here is: whose interests were eventually taken by the commission to constitute the national interest? The answer to this will be sought in the hearings and recommendations of the Borden Commission with respect to three matters: the price and volume of natural gas exports; the construction of pipelines; and the marketing of Canadian crude oil. In addition to its importance as a factor contributing to the formation and eventual operations of the National Energy Board, the commission's record is a valuable indication of the lay of the political landscape after a decade of development in Canada's petroleum industry.

The Commission and Natural Gas Exports: Volume

On the question of gas exports to the United States, it seems convenient to divide the submissions received by the commission into two areas of concern: first, whether gas surplus to Canadian needs should be exported to the United States, and second, how to estimate the size of an exportable surplus.

Several extensive briefs were presented to the commission expressing the concern that Canadian homeowners and commercial and industrial consumers, present and future, should have a guaranteed long-term supply of cheap natural gas. The city of Calgary, for example, proposed the setting aside of nearby fields of "sweet gas" for home and commercial consumption by Calgarians. (Sweet gas is gas that needs very little processing before use, owing to the fact that it contains few contaminants.) Calgary argued, as others have argued since, that the export of such gas meant inevitably that Calgarians and other Canadians would be forced to rely on the development supplies of gas which would either be more remote or would require more processing before use, and would therefore be more costly.[13] The government of Alberta proposed that a thirty-year reserve of gas for Alberta be assured before allowing exports, and the city of Edmonton insisted that increased demand brought about through exports should not be allowed if it entailed higher prices to consumers.[14]

The government of Saskatchewan stated that its policy was to reserve gas for domestic and commercial consumption, and submitted that it would be absurd to export natural gas to the United States at a given price until all accessible Canadian markets had been satisfied at this price. Saskatchewan further suggested that the Canadian government purchase supplies in the ground for later delivery to Canadian markets, thus trying to meet the argument of producers that they must have access to the export market in order to develop new reserves.[15] The Fuel Board of Ontario also felt that Canadian market requirements must be met at the lowest possible price before exports were allowed.[16] B.C. Electric submitted that the National Energy Board should determine the future needs of Canadians and the availability of supplies at a given price to Canadians before authorizing exports.[17] The B.C. Electric brief stressed the role of natural gas supplies as an impetus to Canadian economic development, and cautioned, for example, that if the gas exported was intended for thermal-electric production, Canada would be better off to produce the electricity for export.

This list does not exhaust but does represent fairly the positions taken by those expressing reservations on the desirability of exporting natural gas. The key point in the context of the present discussion is the stress placed by these parties not simply on the long-term availability of natural gas for Canadian consumers, but also on long-term availability at stable prices. However, the parties primarily interested in the production and marketing of natural gas were unanimous in laying stress on availability only—that is, strictly on the size of potential reserves and not on the possible higher cost of future reserves.

The Canadian Petroleum Association (CPA), Westcoast Transmission, Shell Canada, and others continuously stressed the importance of export markets in encouraging the development of reserves.[18] The dynamics of the interdependence between markets and reserve development argued by these parties was that the enlarged demand caused by exports would raise prices and that, in turn, higher revenues to producers would stimulate increased exploration and development.[19] Thus, the producers and the transmission companies were working to shift attention away from the effect of exports on Canadian prices and toward their effect on the expansion of the industry and the development of reserves. If this could be accomplished, the argued interdependence between higher levels of exports and increased rates of exploration and discovery could be used to influence yet another side of the question concerning exports—namely, the formula to be used in estimating the volumes of gas available for export.

There appears to be considerable difficulty in reaching a reasonable estimate of exportable surplus, largely because a variety of formulae can be employed for doing so, as the government of Saskatchewan pointed out in its brief (and as the great variety of estimates submitted to the commission exemplify). Moreover, the key question of relating estimated reserves to estimated Canadian requirements in order to arrive at a figure denoting an exportable surplus is clearly a political rather than a technical matter, since it unquestionably entails a number of judgments in which values play a much more central role than do facts or proven methods, such as the degree to which the meeting of future Canadian requirements should be allowed to depend upon future discoveries.

It is perhaps not surprising, therefore, that toward the end of the hearings of the commission British American Oil submitted a paper entitled "Procedure for Determining Exportable Volumes of Gas," along with a suggestion that the best way to settle the question of reserves was to hold annual meetings on the subject between the CPA and the relevant government agencies.[20] In this and other ways, the hearings of the commission on this question exhibit, in volume after volume, what might be called "the politics of surplus estimation." Since the export of surplus gas was generally accepted to be in the national interest, a conflict arose among interested parties in determining how much gas would ultimately be found and how long it would last. We must turn to the first report of the commission to discover whose estimates, or procedures for arriving at estimates, became the basis for the commission's recommendations.

Recommendation I of the commission's first report endorsed the principle that any gas surplus to Canadian needs should be exported.[21] Moreover, after having projected the supply of, and demand for, natural gas in Canada, the commission suggested that "on any reasonable assumption regarding the growth of reserves in Alberta and British Columbia, there will be a *moderately increasing* volume of gas in excess of Canadian requirements available for export."[22] The commission estimated Alberta's year-end disposable reserves to be 23.8 trillion cubic feet (Tcf) in 1958, increasing to 35.2 Tcf in 1987 by one formula and 52.2 Tcf in 1987 by another formula. The first formula assumed an average annual gas discovery rate of 2 Tcf through to 1970, followed by an annual rate of 1 Tcf to 1987, which the commission claimed was a conservative assumption. The second formula, less conservative, assumed an average annual increase in Alberta proved reserves of 2.5 Tcf, falling off to 2 Tcf annually, approaching 1987. The latter assumption was based on a projection of the average annual increase in Alberta reserves from 1950 to 1958. Why this rate of discovery was expected to continue was not explained by the commission, just as no reasons were given for assuming that, if the higher expectation did not hold true, the more conservative assumption would. However, the key point was the commission's conclusion that "from the evidence available at this time" the higher reserve projection "would be assured if the industry had the added incentive which would be provided by increased export markets," a familiar formulation to anyone who has read the submissions to the commission from the oil- and gas-producing companies.[23]

Turning to Recommendation 2, several points arise from one of the commission's comments on the matter of export licences.[24] The commission began by restating the principle that an application for export should not be permitted if it "would in any way interfere with the supply of the reasonably foreseeable natural gas requirements of those parts of Canada within economic reach of the producing provinces."[25] The commission did not say, however, whether upward pressure on the price of gas to Canadians was to be deemed one of the ways in which an export permit interfered with availability of supply to Canadians. Indeed, the commission was surprisingly silent in some cases and vague in others on the question of prices, especially given the emphasis placed on this matter by many parties represented before the commission.[26] One possible means of ensuring that Canadians would have extended access to gas at low prices, a method supported by at least two municipal governments and one provincial government, was rejected outright: "The Commission believes that, in the administration of export policy, it would be unfair to producers of natural gas to require, at this time, that proven reserves be set aside for all long-term future needs in Canada."[27] That policies recommended as being in the Canadian national interest necessarily entail fairness to producers of natural gas is an assumption which may or may not be valid, but it is one that the commission did not bother to justify.

The Commission and Natural Gas Exports: Prices

When the commission did address the question of the price provisions of an application for an export licence, it had this to say:

> It is necessary to ensure that the minimum export price is fair and reasonable. Where sales to Canadian distributors are involved, the price relationship, between Canadian sales and sales for export, should be such that the Canadian sales will not contribute more than a fair and reasonable proportion of the total return to shareholders on their investment on the gas transmission company.[28]

This statement occurs in the form of a comment rather than a formal recommendation. However, the commission did formally recommend the rescinding of a price regulation (regulation 9) already in force under the Exportation of Power and Fluids and Importation of Gas Act.[29] This regulation read: "The price charged by a licensee for power or gas exported by him shall not be lower than the price at which power or gas, respectively, is supplied by him or his supplier in similar quantities and under similar conditions of sale for consumption in Canada."[30] The Commission's argument for revoking this regulation is also of interest;

> While the Commission believes it understands the result which Regulation 9 was designed to accomplish, nevertheless we have found it most difficult and, indeed, almost impossible to interpret. In the first place, the quantities and conditions of natural gas sales vary greatly as between contracts, so that price comparisons are difficult to make. The usual method of determining appropriate prices is based on a computation of cost of service and there are various methods of allocating certain of these costs to different types and

quantities of sale. Furthermore, the Regulation does not take into account other factors, such as competitive price and value of service, factors which many authorities believe should be taken into account in the setting of the prices.

The Commission believes that, if a National Energy Board inquires into the terms and conditions of each proposed export contract, satisfies itself that the terms are fair and reasonable and in the public interest . . . the objective which the Commission assume were envisaged by Regulation 9 will be achieved.

The logic of the commission's argument against regulation 9 seems somewhat opaque. First, to point out the difficulty of determining whether or not gas is being sold in similar quantities and under similar conditions of sale is one thing; it is quite another to dispense with the principle that *if and when* these similarities can be established, the export price shall not be lower than the Canadian price, which is the only reasonable interpretation of regulation 9. Second, while there is undoubtedly truth in the commission's claim that the problem of determining the terms and conditions of export contracts is highly complex, there is no obvious reason why the problem should be any less complex when the licensing authority is judging a contract by the commission's standard of "fair and reasonable and in the public interest" than when it is judging the contract by regulation 9's standard of an equal or lower price for Canadian customers. Third, one wonders how the commission, which found regulation 9 difficult to interpret, could seriously propose that the National Energy Board should extract unambiguous meaning from the phrase "fair and reasonable and in the public interest."

In the light of later developments, we are left in an ambiguous position with respect to the export-pricing arrangements recommended by the commission. On the one hand, it appears from their comment that they were conscious of the importance of such factors as competitive prices and value of service ("opportunity prices"), the need for escalation clauses in export contracts, and the effect of exports on the market for byproducts. On the other hand, the wording of most of their actual recommendations in these matters—wording which often found its way into the National Energy Board Act—was so vague that they could not ensure that these considerations would be effectively applied to the decisions of the board.

The Commission and Pipeline Construction

The Borden Commission recommended that the jurisdiction over pipelines be divided between the regulation of tolls, rates, or tariffs, which would remain with the Board of Transport Commissioners, and the regulation of the construction of new pipelines, which would become the responsibility of the new NEB.[31] The commission devoted considerable attention to the question of the initiation of new pipeline projects, primarily by virtue of its investigation into the financing and the actual or proposed export arrangements connected with both the Westcoast Transmission and Trans-Canada Pipeline projects.

In its recommendations concerning the issuance of licences and certificates of public convenience by a new regulatory agency, the commission added to the considerations already discussed with respect to export licences "the advisability of encouraging the development in Canada of processing industries relating to energy and sources of energy as distinct from the export of unprocessed raw materials."[32] It further recommended that the NEB also be required to take into account, first, the economic feasibility of the project and whether or not the project is in the national interest and, second, the financial structure, ownership, financing, engineering, and construction plans involved, as well as the opportunity for Canadians to participate in the financing, engineering and construction of the project.[33]

It is noteworthy that the commission failed to address explicitly two issues that had frequently arisen in connection with the construction of pipelines in Canada, namely, the routing of pipelines from Canadian sources to Canadian markets completely within Canadian territory, and the desirability of projects constructed solely for export service. These issues might have been expected to receive some attention, since they were raised in Parliament during the earliest pipeline decisions and were important aspects of the two projects the commission examined at some length. The Trans-Canada project would never have become the centre of controversy it was had it not been for the insistence on an exclusively Canadian route; and the advantages and disadvantages of sharing costs between Canadian and export markets were probably the central points of contention in the Westcoast controversy.

The Commission and the Alberta-to-Montreal Oil Pipeline

The commission did consider one major pipeline proposal with broad national implications—the proposal to serve the Montreal market with Alberta crude oil by means of a pipeline built exclusively within Canada. The proposal was made by Home Oil Company Limited, with a group of other companies supporting the project. It was proposed to construct a 30-inch diameter pipeline running from Edmonton to Montreal, with the promise of delivering 200,000 barrels per day to Montreal by 1960 and 320,000 barrels per day by 1965, in order to provide Alberta's crude oil producers a guaranteed market.[34]

Numerous other benefits to Canada were claimed to follow from this plan, but they will not be discussed here except to say that two obvious implications of the scheme were the cutting back of Canada's rather high rate of oil imports and, at the same time, the lessening of the Canadian industry's dependence on the United States market. It is important to note, however, that the Home Oil representative seemed to suggest that the overriding concern of his company and its associates was to obtain assured markets; and that any market, such as the west coast or the midwestern United States, would suffice if these markets could assume the permanence of the Montreal market.[35] Another important point in the presentation was the expressed recognition that the financial success of the pipeline could be assured only on the basis of guarantees from the Montreal refiners to use the full capacity of the line, that is, their willingness to accept the crude oil carried by the line in place of the oil they obtained at that time from their overseas sources.[36]

The opposition to the Home Oil proposal was unanimous among the large foreign-controlled corporations, specifically Imperial, Shell, McColl-Frontenac, British American, Standard Oil of California, Sun, and Canadian Petrofina.[37] Imperial Oil led the opposition, claiming that government protection of the Montreal market for Canadian crude was mandatory if the pipeline was to be feasible, and that such protection would entail both higher prices for Canadian consumers and greater government control of the petroleum industry, both of which were rejected as undesirable, especially the latter. However, Interprovincial Pipelines, while questioning the advisability of a new crude oil pipeline to serve the Montreal market from Western Canada, submitted a counter-proposal to upgrade its present crude oil transmission facilities as an alternative to the Home Oil proposals.[38] When Imperial was asked whether, as a refiner, it would sign throughput agreements with such a pipeline if one were undertaken, the company replied that it would reluctantly fall into line with any requirements laid down by the government.

All the companies listed above took nearly identical positions on the question. The principle of bringing Canadian crude to the Montreal market was rejected; a reluctance to sign throughput agreements with the proposed pipe line was expressed by most and a refusal to do so was promised by at least one; alternative market opportunities for Canadian crude, such as the north central and Pacific states, were invariably pointed out; and governmental interference in Canada's crude oil market was generally condemned.[39] Several companies expressed a preference for a continental energy policy, whereby all North American markets would be served by the most economical sources of supply. Since all the companies mentioned so far were subsidiaries of integrated international petroleum enterprises, with interests and affiliates in production, refining, shipping, and marketing in numerous countries, the similarities in the substance of their proposals can be readily accounted for. It should be noted that both the American government, for political reasons, and Imperial's parent company, for economic reasons, were concerned about the impact of United States import restrictions on Venezuela, then a major supplier to eastern Canada, and would not have favored the loss of the Montreal market to Venezuelan subsidiaries of the international companies, the largest of which was an affiliate of Imperial. Other companies without such worldwide affiliations, some of them Canadian-controlled, also took a stand opposing the proposed pipeline and promoted similar alternatives. Canadian Oil and the Bailey Selbourne Group are examples.[40] However, no Canadian subsidiaries of international oil companies *supported* the plan.

Still, it is hard to sustain the argument that the Montreal pipeline issue pitted the Canadian interest or Canadian nationalism against the power of the American government and American firms. While these factors were unquestionably present, there is little evidence of Canadian interest in the scheme except on the part of the Canadian independent firms, and even then only if reliable American markets could not be obtained. Little support for the policy emanated from central Canada: Ontario already received Alberta oil, and no representatives from Quebec even addressed the issue in Parliament. This lack of support for the idea in Quebec was probably a reflection of the fact that Alberta oil would be more costly than Venezuelan oil delivered to Montreal.[41] No one seemed impressed with the security advantages of self-sufficiency for Canada, despite the Suez crisis only two years before. In sum, the only Canadians with a strong interest in the movement of Alberta oil to Montreal were the independent Canadian oil producers, and even they had more to gain from seeking a reduction in the American restrictions against imports from Canada.

The recommendation of the commission was flatly to reject the proposal submitted by the Home Oil Company and associated companies.[42] The option staunchly defended by the international petroleum enterprises formed the basis of the commission's recommended policy: the Montreal market was to be reserved for overseas sources of supply, while Canadian production was to be increased by penetrating accessible United States markets and by market expansion in Ontario and Vancouver. While one cannot argue that the recommendations of the commission required no changes in the operations of the major companies operating in Canada, it is equally obvious that the recommendations reflect precisely the expressed preferences of these firms as to the changes they would have to make.

Ultimately, it can only be concluded that the commission succumbed to the economic rationality of a continental as opposed to a national approach to the problems of oil production and marketing. Referring to a recent exemption of Canadian crude from United States import restrictions, the commission stated:

> While we realize that the many possibilities, problems and implications may not have been fully reviewed as yet, this exemption could be the first step leading towards the development of a continental policy with respect to crude oil under which Canadian and United States crudes would be freely used in the refinery areas on the North American continent, supplemented by such imports of crude as might be necessary to augment any shortage of supply from North American sources. We mention the possibility of a continental policy not because we believe it can be developed in the immediate future but because we feel that care should be taken to ensure that Canada, by its actions and commitments now, does not jeopardize the subsequent possible development of such a policy.[43]

Three pages later, the commission explained its recommendation that no import restrictions be imposed to secure the Montreal market with the claim that such a decision, "if made before the potentialities of United States markets were fully exploited, would, among other things, seriously impair Canada's position vis-à-vis existing United States import restrictions and might jeopardize the development of a Continental energy policy."

The commission went on to set down proposed guidelines for the National Energy Board to consult in exercising its authority to grant licences for the importation of crude oil. The commission felt that the board's control over import licences, plus its control over the flow of Canadian crude by means of its authority over pipelines, would be sufficient power to enable the NEB to shape the development of the Canadian petroleum industry. But the commission admitted that "we have not attempted to set out the details with respect to a licencing system because we realize that exceptions might be required for certain types of crudes and that problems of a technical nature may be involved." The report continued:

> We are of the view that the oil industry itself is able to supply any information and to assist in the resolution of whatever administrative difficulties may arise in putting into effect such licencing procedure. The National Energy Board, as a permanent body of the Government of Canada, provides a forum where the industry can discuss its problems at the Canadian government level. What is perhaps of more importance, this Board as an agent of the Government can and should keep in close touch at all times with the industry, in all its phases, and with all its problems, as these have a bearing upon the prosperity of the Canadian economy and of the industry itself. Consequently, we believe that the problems involved in such licencing procedure should be resolved through discussions between the Board and the industry itself.[44]

The commission thus endorsed the principle that the Canadian national interest in petroleum matters should be defined by means of government-industry consultations. As we shall see, the National Energy Board did indeed undertake close consultation with the representatives of the industry and came to rely on them for data and judgments concerning the issue of gas exports as well as the administration of the National Oil Policy, which was formally announced a short time later.

Before leaving the commission's deliberations and recommendations concerning the Alberta-to-Montreal pipeline, it is worth recording that the project was given considerable attention as well in the House of Commons. Supporters of the venture tended to chide the international oil companies in Canada for their reluctance to open the Montreal market to Canadian production, but it is also striking that most MPs who spoke on the subject seemed to prefer a solution to the plight of Alberta producers in the form of increased access to the American market, to be achieved by means of a relaxation of the U.S. quota on imports of Canadian crude. For instance, Hazen Argue of the CCF used the occasion of a debate on the incorporation of Stanmount Pipeline Company to criticize Imperial Oil for its opposition to the Alberta-to-Montreal project. The stand taken by Imperial Oil, he remarked, "has been a narrow one and has been governed solely by the financial interests of Imperial Oil, and it is an endeavour by this international oil cartel to maintain the Montreal markets for oil from Imperial Oil and its associates."[45] Argue, however, also referred to the prospect of reductions in United States import quotas. He hoped that the U.S. would not take this action, and that "in a short time, restrictions on Canadian oil entering that market would be removed." The CCF attitude toward the Montreal pipeline was thus contingent upon the crude oil import restrictions of the United States government. In lieu of American markets, the CCF seemed ready to support the independent oil producers in gaining access to the Montreal market.

Members of the two major parties also seemed to give priority to gaining access to the U.S. market. It is true that Arthur Smith, a Conservative from Alberta, supported the independent producers in their efforts to move Canadian crude to eastern markets, stating that he would "continue to support moving more oil out of western Canada to eastern Canada because of the absolute essential necessity of providing assistance to the independent operators."[46] This position was not supported in earnest by the Diefenbaker government. In reply to a question regarding the U.S. oil import quotas, Prime Minister Diefenbaker stated that

> the government will seize every opportunity to obtain unimpeded access to the United States market for Canadian oil. In this or in other ways, which may be necessary, it is our determination to safeguard the interests of the Canadian oil industry.[47]

In another statement, Diefenbaker pointed to his government's success in having the import restrictions eased, allowing exports of Canadian crude to rise from 15,000 barrels per day to 25,000 barrels per day. The prime minister regarded this as an "important and valuable concession."[48] He was evasive, however, on the government's position on the desirability of the Alberta-to-Montreal pipeline as an alternative to the potential of the United States market. Replying to a question by the CCF concerning the continuance of the United States import program and the possibility of construction of a Montreal pipeline if Canada's exports should be reduced, Diefenbaker replied only that the "course of action to be followed by the government will be determined in the light of the fullest study."[49]

It would appear that the Conservative government under Diefenbaker preferred to pursue the removal of American import quotas rather than to give a firm commitment to the construction of a pipeline to the Montreal refineries in order to provide a wider Canadian market for Canadian crude. This position was remarkably similar to that of the opposition Liberals under Lester Pearson. Pearson clearly and explicitly endorsed a continentalist philosophy based on economic rationality and pragmatism. In a frank statement, he concluded that

> if defense is to be considered on a continental basis, then resources and materials for continental defense must also be based on a continental basis.[50]

In order to facilitate continental resource sharing, Pearson suggested later in the same speech that a

> joint materials board especially concerning base metals, oil and supplies of strategic materials could at least be considered by both governments on a continental basis.[51]

Thus, both major parties avoided making precise statements concerning the Montreal pipeline project. The Liberals appeared quite satisfied to pursue a program of joint Canadian-American resource sharing, while attacking the government for not pursuing Canadian exemption from the United States import program with sufficient vigour. Diefenbaker similarly approached the Montreal pipeline issue in the context of American import quotas: like Pearson, he regarded access to United States markets as being more important than the construction of Montreal pipeline and a reduction of Canada's dependence on overseas oil.

Results of the Borden Commission's Recommendations

The reports and recommendations of the Borden Commission experienced a reasonably high rate of government implementation. With respect to matters relating to exports and pipeline construction in general, several commission recommendations can be seen behind certain sections of the bill creating the NEB that was introduced by the Diefenbaker government. Moreover, the commission's lack of enthusiasm for the Alberta-to-Montreal pipeline, and its justification for that attitude, can clearly be seen in the National Oil Policy announced on February 1, 1961, by the same government. But beyond these instances of influence upon legislation and policy, the commission also set a precedent for the NEB with respect to procedures and methods to be employed in exercising the jurisdiction over trade and transportation of oil and gas that it was soon to inherit.

With respect, first, to natural gas, the National Energy Board Act attached fewer and less specific conditions than did the commission to the export of natural gas and the construction of pipelines for that purpose. In particular, there was nothing in the act that expressly reflected the commission's concern that exports should not be allowed to the detriment of potential industrial development in different parts of Canada; there was nothing that incorporated the commission's criteria according to which the price of natural gas exports might be deemed "fair and reasonable and in the public interest," such as cost of service, competitive prices, and value of service; and there was nothing that reflected its recognition of problems associated with the marketing of byproducts of natural gas.

However, one recommendation of the Borden Commission was translated more or less directly into the National Energy Board Act, namely that, "having regard to the proven reserves of natural gas in Canada and to trends in the discovery and growth of reserves, the export of natural gas, which may from time to time be surplus to the reasonably foreseeable requirements of Canada, be permitted under licence."[52] The commission argued for this recommendation in a manner which also illustrated how it might be made operational. It made, first, a thirty-year projection of supply and demand in the Canadian market and, second, an estimate of the potential exportable surplus that might be available over the same period, depending upon the rate of gas discovery and development. It seemed probable to the commission that the upper limit of the reserve projection "would be assured if the industry had the added incentive which would be provided by increased export markets."[53]

There is a circularity to this reasoning which has persisted in export policy ever since. It is born of the producers' argument that the best protection with regard to future supply is the export of gas. The circularity consists in the fact that the volumes of gas that are counted upon to justify the export of gas (that is, the volumes of gas necessary to protect future increases in demand in the Canadian market) are allowed to depend upon the export of gas for their development. Whether the rate of discovery and development of reserves has, in fact, any relationship whatsoever to the rate of export of gas is a question open to empirical investigation. But a problem can arise even if the new discoveries are so generated. The gas to be exported and the gas whose expected development is supposed to justify exports are, of course, different lots of gas; the latter may only become available at higher cost and in more inaccessible locations than the exported gas it is counted on to replace. Hence, it is possible that by making exports allowable on the basis of trends in discovery, proven reserves at current prices will be exported while the protection of Canadian requirements will depend upon future reserves at higher prices. The commission seemed to be sensitive to the dangers in this. As related above, it was prepared to concede that future Canadian requirements should be protected by reserve growth trends on the basis that "it would be unfair to the producers to require, at this time, that proven reserves be set aside for all long-term future needs in Canada," but it also urged that "the Government of Canada should require satisfactory evidence in respect of reserve growth trends and evidence that the supplies of natural gas, expected to become available by reason of the trends, are suitably located for transmission to Canadian markets."[54] As chapter 6 will show, the government and the NEB seem to have taken up only the first of these injunctions from the commission.

If anything, the act was even less specific about the conditions which must be met before exports would be allowed. A partial explanation for this may be the attitude of the government and the official opposition toward gas exports. The government and the Liberal opposition, while they had differences over some aspects of the bill, showed a common concern that as few obstacles as possible be placed in the way of the industry, whereas the CCF cautioned that what might be best for the industry, especially if it meant high volumes of exports at less than full value, was not necessarily in the public interest.[55]

A further example of this lenient attitude is provided by an occasion on which the government sought to amend the original bill by adding the phrase "having regard to the trends in the discovery of gas in Canada" to clause 83. The Liberals supported this amendment after Pearson had taken pains to establish through questions to the minister that the amendment was designed to "ensure that there shall not be an unnecessarily restrictive policy applied by the Board" and that foreseeable requirements for use in Canada "should be interpreted in the light of potential production."[56] In other words, it sounded as though the main concern of the Liberals was to ensure that the board did not become a significant obstacle in the way of gas exports. However, it must be noted that the Liberals did seek, unsuccessfully, to amend clause 83(b) to read, as did the old regulation 9, to the effect that export prices must not be lower than the price for similar Canadian sales, an amendment that would seem to owe a debt to memories of the scandal arising out of the first Westcoast export contract.

A comparable degree of agreement existed between the two major parties when it came to the disposition of Canadian oil and the adoption of the famous Ottawa Valley Line. After considering the Borden Commission findings for well over a year, and reflecting the kind of thinking on this issue which both Diefenbaker and Pearson had voiced earlier, the Conservative government announced its National Oil Policy. Following the recommendation of the Borden Commission, the Ottawa Valley became an east-west dividing point for the marketing of Canadian crude. The Ontario market and points west were to consume Alberta crude, while all markets east of the Ottawa Valley would remain dependent on foreign sources. A pipeline to Montreal was not seen as necessary. As Gordon Churchill, Minister of Trade and Commerce, announced, the

> increased production would result through increased use of Canadian oil in domestic markets west of the Ottawa Valley and by the expansion of export sales largely in existing markets.[57]

He further stated that:

> increases in exports, which is integral to the government's program, is wholly consistent with the growth of sales of Canadian oil contemplated when exemption from U.S. import quotas is established.

The Liberal response to this announcement was as innocuous and uncontentious as one might expect, given their leader's earlier expression of enthusiasm for continental trading arrangements for natural resources in general and his view that "in a well-ordered continental society" the great proportion of our oil production should be going to U.S. markets.[58] The announcement of the NOP simply prompted the observation that the government statement amounted to nothing more than the policy recommended by the Borden Commission; the Liberals criticized the government merely for the amount of time it had taken to act on that recommendation.[59] The CCF supported the policy "to the extent that [it] will encourage production by independent oil companies." Events were to prove, of course, that this was a rather empty and unrealistic wish, as many of the independents who had looked to the Montreal venture as a solution to their problems gradually sold out to the majors after 1961. Indeed, the adoption of the very name "National Oil Policy," while perhaps inevitable given a prime minister of the day who self-consciously assumed the mantle of Sir John A. Macdonald, seems scarcely legitimate, given what the concept of a national policy had generally been accepted to mean in the past; for the policy rejected the idea of serving both Canada's major fuel markets with Canadian-produced oil, and promised instead merely to prevent imported oil from penetrating further into the Ontario market than it was threatening to do. As such, the policy is reminiscent of the content and purpose of the assistance afforded Nova Scotia coal in the 1920s and 1930s. But as we have seen, in those days such a policy did not warrant the label "National Fuel Policy," for it left a large portion of the Canadian market dependent upon foreign supplies of fuel.

The implementation of the recommendations of the Borden Commission brings us to the early 1960s. From this point forward, the matters under review in this study became the responsibility of the National Energy Board, which not only regulated gas exports and the construction of pipelines, but administered the provisions of the NOP. Chapter 6 will provide a review of the NEB's exercise of these powers between 1960 and 1971, when a transformation of the world petroleum market overturned all the conditions upon which policies up to that point had been predicated, and when pressures on the federal government grew for an early decision in favour of the Mackenzie Valley Pipeline. We shall see that by 1971 (the year of the last major decision to be taken by the board prior to the Mackenzie Valley issue) the board had done nothing to expand the Canadian consumption of Canadian fuels much beyond levels already inherent in the industrial and policy structures it first encountered in 1960 and, if anything, helped to deepen those continental arrangements of which a northern gas pipeline was to be the culmination, if not the planned objective.

6 Natural Gas and the National Interest: National Energy Board Decisions from 1960 to 1971

THROUGHOUT most of the 1960s, the central thrust of Canadian energy policy was to encourage the export of Canadian oil and natural gas to the United States, and there was little public opposition to the decisions taken in promotion of this objective until late in the decade. The export orientation of Canadian policy makers and the lack of public controversy over export sales are both explained by the fact that, throughout the decade, cheap and apparently reliable supplies of oil were available from foreign sources, so that the desire on the part of western oil and gas producers to expand the industry by securing a larger share of the American market did not conflict with any strongly asserted desire on the part of central Canadian consumers to reserve Canadian fuel supplies for their own present or future use. Moreover, Canadian oil and gas reserves were on the increase during this decade, and there seemed little reason to fear that a rising rate of exports was in conflict with any national priorities. In short, the condition of the world petroleum market at the time was more conducive to an international than a national pattern of trade in oil and gas. Consequently, by virtue of the recently established National Oil Policy, roughly half of the Canadian market for oil was open to imported crude oil, while the other half was reserved for the (slightly higher-priced) Alberta product; and by virtue of decisions taken by the National Energy Board during the 1960s, roughly half of Canadian production of natural gas would be sold in Canada, while the remaining half was exported to the American midwest and northern Pacific coast. Finally, throughout the decade, the NEB and the federal government were engaged in a persistent endeavour to maintain or expand the amount of Canadian oil exported to the United States.

But the primary responsibility of the National Energy Board, following the recommendations of the Borden Commission, was to regulate the volume and price of natural gas exported from Canada as well as the construction of new pipelines for the purpose of international and interprovincial transmission of Canadian gas supplies. In exercising its responsibility to regulate the export of gas from Canada, the board was required to make three types of decisions when an application was made for an export licence. First, the board had to determine whether a surplus of gas was available; that is, whether Canada's established reserves of natural gas were sufficiently in excess of Canadian requirements to cover the volume of exports applied for at any given time. Second, the board had to decide whether the price to be charged by the applicants for the gas they would be selling to their American clients was "just and reasonable." Third, the board had to decide whether the plans for any new or expanded pipeline facilities that might be required by the export proposal were acceptable. Between 1960 (the year of the first export decisions taken by the board) and 1971 (the year of its last significant export decisions prior to the OPEC price increase of 1973) the board made dozens of decisions in the areas of surplus determination, export prices, and pipeline construction. Taken together, board decisions during this period reveal an "export orientation" on the board's part, by which is meant a primary concern to sustain and, if possible, promote the future growth of the American market for Canadian natural gas.

In the area of surplus determination, the export orientation of the board was evident in its consistent willingness to commit to export markets all of Canada's established gas reserves in excess of Canada's currently contracted requirements or, in other words, its repeated refusal to set aside established reserves of Canadian gas for the protection of increases in Canadian requirements, a policy which left the future expansion of the Canadian gas market to depend on the future development of new gas supplies for its protection. In the area of export prices, the export orientation of the board was clearly shown in its refusal to insist on export prices that reflected the market value of the gas in the American markets in which it was ultimately sold, as measured by the cost of alternative fuels in those markets. In other words, the board revealed that it was more concerned with expanding the volume of export sales than with realizing the full value of those sales. Finally, in the area of pipeline construction, the board made several decisions to approve new or expanded pipeline facilities that were explicitly justified, at least in part, with reference to the favourable impact they would have on the future marketability of Canadian oil and gas in the United States.

In all three of these areas of jurisdiction, board policies both reflected the concern to promote a continental marketing arrangement for Canadian oil and gas espoused by the Liberal party and the Borden Commission in the late 1950s, and anticipated the justification for the approval of the "pre-build" portion of the Alaska Highway pipeline in 1980. The Canadian oil and gas industry must grow; if it is to grow, it must not sit on unsold reserves of oil and natural gas; if it is not to sit on such unsold reserves, they must be exported, whether or not parts of Canada have or will have need of them: so runs the consistent (if at times only implicit) rationale for expanding Canadian oil and gas exports in preference to expanding (or prolonging) the Canadian consumption of Canadian fuels. However, the NEB's responsibility in these matters must be evaluated in the light of the broader economic and political circumstances of the 1960s. A high rate of exports did serve the interests of the oil and gas industry, but it also served the interests of the governments of the producing provinces—who by then were confident that their gas reserves were comfortably in excess of their own future requirements—in promoting provincial development. In turn, this provincial economic growth also served the central Canadian provinces, which enjoyed a growing market for their manufactures, while the gas exports contributing to this western development did not appear to threaten their energy supply, given the relatively low cost and apparent reliability of world oil supplies.

However, at the close of the decade, parties concerned with the Canadian market did begin to perceive some threat to their interests from the expansion of export sales: future discoveries of gas in Canada, upon which the protection of Canada's future requirements had come to depend, seemed more and more likely to come only from the more remote and more costly gas supply regions of the country. This altered the balance of political forces on the question of gas exports and revived some conflicts over natural gas exports that both recalled the debate of the mid-1950s and anticipated some of the developments of the 1970s. The NEB export decisions of 1970 and 1971 became the focus of these emerging conflicts.

Gas Exports: Determination of Surplus

The two criteria stipulated in the National Energy Board Act for the acceptability of an application for the export of natural gas were, first, that the volumes to be exported were surplus to Canadian requirements and, second, that the price at which the gas was to be sold at the border was just and reasonable in relation to the public interest.[1] The board evolved after its inception an elaborate and complex procedure to give effect to these provisions of the act, but essentially the procedure was to determine the size of an exportable surplus without reference to the terms of individual export applications, and then, only if there was a surplus to be exported, to consider which export proposals would be accepted or rejected on the ground of price. In practice, the first step was all-important, for as shown below, throughout the 1960s the board and the Cabinet without exception approved exports to the full amount of the exportable surplus that, from time to time, had been determined to exist.

The rate at which Canada exported natural gas to the United States, therefore, depended essentially on the apparently technical matter of how the NEB operationalized the basic formula: exportable surplus equals established reserves minus Canadian requirements (ES = ER – CR). There is no question that in its detailed procedures this was a technical matter, but it is also true that the board's use of data, methods, and formulae in arriving at such estimates was, and indeed had to be, also a matter of political judgment. These judgments concerned such questions as the span of time over which Canadian requirements were to be estimated; the method of estimating these requirements and taking them into account; the definition of what was to constitute established reserves; whether or to what extent future Canadian requirements were to be protected by established reserves only or by established reserves plus some portion of future reserves; and the basis for projecting the rate of growth of reserves. In deciding these questions, the board relied on several sources. First, the government of Alberta and its (then) Oil and Gas Conservation Board provided the National Energy Board with several operating principles and formulae. Second, the Borden Commission had established several procedures, formulae, and precedents in grappling with similar issues. Third, over the years, firms in the petroleum and pipeline industries, either individually or collectively through their trade associations, made significant contributions to the definition of these procedures.[2]

During its early years of operation, the board was able to set and apply these procedures in a relatively uncontentious atmosphere, since none of the parties to its proceedings—producers, export-oriented pipeline companies, Canadian-oriented pipeline companies, producing provincial governments, and Canadian distributors of natural gas—raised any serious challenge to the applications made by any of the others. However, this condition did not appear as though it would continue to hold in 1970, as interests oriented to the Canadian market began to perceive that the increased exports to be approved in that year might threaten the availability and price of gas in Canada. This gave rise to a revival of the debate over the appropriate procedures for determining exportable surpluses of gas that had gone on before the Borden Commission ten years earlier, and to the repetition of the old arguments concerning the most effective means of protecting Canada's future requirements. While it is true that in the end the National Energy Board declined to approve the full amount of the exports applied for on this occasion, the export orientation of board decisions taken up to and including the 1970 decision was evident in the fact that the board did, on this occasion, liberalize its export formula in favour of export sales despite the objections raised to such a move by parties interested in the

Canadian market. Even more significantly, the board acknowledged that even its former procedures had tended to afford more secure protection to export than to Canadian markets.

In November 1969, the board commenced its hearings to consider applications from several firms for export licences totalling 9.5 Tcf. Prior to the opening of this hearing, the board, in the words of the *Annual Report*,

> decided that it must, in view of the amount of gas proposed to be exported, re-examine the principles on which it has previously considered such matters as Canadian gas requirements; indicated gas reserves, deliverability and projected trends in rates of discovery; and the surplus remaining after making due allowance for the reasonably foreseeable requirements for gas use in Canada . . . The Board invited evidence and advice from all other interested parties on these and related questions.[3]

During these hearings, the question of the board's calculation of surplus received attention from a variety of interested parties, and the terms in which they addressed this question leave no doubt that what they perceived to be at stake was the degree of protection of supply afforded Canadian markets or, conversely, the amount of gas available for export. For example, the Canadian Petroleum Association stressed as a very important consideration "that current and future surplus be calculated in such a manner as will minimize to the greatest degree possible the risk of contractable reserves having to be set aside in excess of those volumes of gas for which the purchasers of Canadian requirements are actually prepared to contract for [sic]."[4] The CPA also suggested that the projection of current Canadian requirements beyond a ten-year period could not "reasonably meet the test of forseeability as required under section 83 of the Act," and recommended several changes in the method of calculating supply aimed at enlarging the supply estimate.[5] The Independent Petroleum Association of Canada stated its agreement with the idea "that we now have sufficient information about the potential of Canada to justify a liberalization of the formula for determination of surplus."[6] Specifically, the IPAC recommended that the board take into account 50 percent of the reserves then classed as beyond economic reach, "and include in that category as discoveries are made the initial estimates of probable reserves for new discoveries."[7] There was, in addition, almost complete unanimity on these points among producers who dealt with them in individual submissions, such as Gulf, Shell, Amoco, Amerada Hess, and Mobil.[8] The one exception was Canadian Fina, which expressed support for the current criteria and method of calculating surplus.[9]

Opposition to changes in the board's established procedures and support for a conservative attitude on exports was voiced most strongly by Canadian distributors and consumers of natural gas; that is, B.C. Hydro, Gaz Metropolitain, Consumers' Gas, and Union Gas.[10] They were joined in this opposition by all of the provincial governments represented, except Alberta. Of the four exporting companies, Trans-Canada and Westcoast which, unlike Alberta and Southern, serve Canadian as well as American markets, did suggest some changes in methods, but generally argued for a conservative approach to determination of surplus.[11] It was argued by some of these parties that the proposed departures from established procedures would have the effect of freeing volumes of gas for export that had hitherto been reserved for the protection of Canadian markets and of increasing the dependence of Canadian consumers on future and more costly supplies. It was, for example, Quebec's view that

> there is no need to accept the suggestion made to include all or part of the reserves beyond economic reach or those deferred for conservation purposes in order to establish available reserves. In fact, these reserves, whether proven or not, are not available to the market, and their inclusion in available reserves could only increase artificially, to the detriment of the Canadian consumer, the surplus for export purposes.[12]

These sentiments were echoed by Ontario. Both provinces pointed out that only by revising its formula for determining surplus could the board grant all the exports sought. The danger which most of the opponents saw in doing so was well expressed by B.C. Hydro: "We are opposed to any policy of committing all known reserves to the support of export contracts, as this would mean that Canadian customers would be called upon to pay a disproportionate share of future discovery and development costs."[13]

The clear concern of all these parties was to protect Canadian markets in terms of the price and availability of gas in the face of rapidly increasing demand in American markets. Their general desire was that this protection be afforded, so far as was feasible, out of established reserves; their general fear was that the protection of their sources of supply would be allowed to depend too heavily upon future discoveries at higher prices. None argued that gas would be unavailable in the future in the sense that the required gas was not there or could not be developed; but they did argue that it could only be made available at price levels which they would find harmful. They opposed a liberalization of the surplus calculation because to do so would be to increase the amount of established reserves committed to export markets, which in turn would made future Canadian requirements more dependent upon "trend gas," that is, gas yet to be discovered and developed. The danger they saw in this was that a condition of trend gas was higher gas prices.

Those advocating changes in the board's methods presented in a different light the relationship between exports, Canadian requirements, and future reserves. According to these parties, the best protection for future Canadian requirements was an expanding industry, and the best assurance of an expanding industry was a rapidly expanding market, that is, a rising rate of exports. It will be remembered that this argument had also been placed before the Borden Commission, and by many of the same parties. As the spokesman for Amoco put it, "new discoveries to be made afford to the Canadian consumer the best possible protection for his future gas needs."[14] However, it was argued that these new discoveries would occur at a rate reflecting the incentive available in the form of rapidly expanding markets, which came down to exports. To quote Dome: "If the export of gas in Canada is discouraged or restrained, we run the risk the vast potential resource of the Canadian north will never be utilized and will therefore be lost to the Canadian consumer."[15]

This basic argument that future Canadian requirements are better protected by liberal than by conservative export policies (and that, consequently, the board should alter its current practices), was repeated by the CPA, the IPAC, Amoco, Banff, Amarada Hess, Dome, Shell, and Mobil.[16] Canadian Fina adduced the same argument, but in a context other than that of the surplus calculation.[17] In addition to these producers and producers' associations, Alberta and Southern and the government of Alberta advanced the same argument.[18] Many of these parties admitted that their argument assumed both the development of northern or offshore reserves of gas and higher well-head prices. Some of them, of course, were also already active in northern exploration. For example, Amoco pointed out that the cost of exploration and production in the more remote areas of Canada "is extremely high" and that, as a result,

> it will be necessary to establish higher gas prices as well as other economic incentives to the benefit of gas producers. Unless such incentives are made available, it will be extremely difficult for the producers to justify the acceleration of drilling programs to the extent necessary to fulfill the indicated rapidly growing future market requirements.[19]

In essence, this came down to an argument that the board should provide for future Canadian requirements not by denying export applications, but by approving them. However, if Canadian access to future supply was thus dependent upon increased exports, then Canadian consumers would pay the higher prices of remote reserves or go without gas in the future. In this sense, the export of Canadian gas once again involved a conflict between the interests of producers and consumers.

The truth of these propositions is hard to assess conclusively. However, the board itself recognized their force and expressed a concern to avoid the potential harm to the interests of Canadian consumers they might entail. For instance, in announcing that it was prepared to shorten the normal twenty-year term of new export licences as a means of preserving more gas for Canadian use, the board had this to say:

> This general approach would have the merit of diminishing in some degree the force of the argument, advanced by some Canadian distribution companies and some provinces, that the granting of export licences for long terms, with full protection as to supply, tends to dedicate an undue proportion of presently available reserves to export markets, leaving Canadian requirements beyond the relatively short term to be met out of gas yet to be discovered, and probably to come at higher cost from more remote areas . . . If . . . increments of export throughput were licenced for relatively short terms, United States and Canadian customers would share more equitably in whatever may be the costs of future increments of supply to be committed to Canadian and export markets.[20]

Moreover, the possibility that future Canadian requirements would have to depend for protection upon the expansion of the industry was apparently real to the board:

> The distribution companies will not be adequately serving their own interests and those of their customers if they fail to contribute their share to the incentive for that increased rate of discovery which is essential if the Canadian gas producing industry is to continue to develop. They cannot safely assume the limitation of exports will by itself ensure that adequate supplies are available at reasonable prices.[21]

These two statements by the board seem to have taken the discussion some distance from the specific question of the formula and procedures for determining whether a surplus of gas was available for export. That, however, is precisely the point. The apparently technical problem of making and utilizing such estimates as Canadian requirements, Canadian reserves, and trends in discovery was inextricable from the political question of how to distribute the various costs and benefits associated with bringing Canadian gas to Canadian and American markets, and, in any case, was inherently incapable of resolving the question, "How much protection for Canada's requirements is enough?" Lurking behind the technical discussions of the size of the available surplus of gas in the West was the issue of the price of gas in Canada, which would in turn affect the ultimate size of the market for gas in Canada. On this issue, the board came down on the side of exports, as it generally had over the past decade. In doing so, the board also accepted, in principle at least, that exports from established, low-cost areas should be permitted despite the fact that this would hasten the arrival of the day when Canadians would have to gain access to remote and much more costly reserves.

Gas Exports: Price

Following the OPEC price hike of 1973 and an increase in the price of both oil and natural gas in Canada, the government usurped the NEB's previous role in determining whether the price provisions of particular export applications were acceptable or not. It did this simply by setting the border price at a uniform level applicable to all export sales to the United States. As a result, the board's earlier procedures and principles governing the approval or otherwise of export applications have little direct relevance today. They are, however, of some importance to this study, for the board's record in these matters reveals that it rarely insisted that Canadian gas exports be priced at a level which would achieve their full value in the American markets they served. This would seem to stand as further evidence of the board's favourable disposition toward exporting any and all gas determined to be surplus to Canadian requirements, since, as noted above, the board always approved export applications to the full extent of the existing surplus, and on at least two occasions the board approved export applications that it explicitly acknowledged had failed to meet one of the three conditions which it had earlier stipulated must be met if the export price was to be ruled acceptable. The board's regulation of export prices during the 1960s reveals that it placed a higher priority on a high rate of exports than it did on receiving the full value of Canadian natural gas, and that exporting gas at less than full value had a higher priority than holding established reserves for future or expanded Canadian use.

Any discussion of the NEB's past decisions to approve the price applications for the export of gas is complicated by the board's lack of consistency in applying its own criteria. The difference between the board's enunciation of established criteria and principles and its application of them to individual cases was nowhere greater than with respect to price. Nevertheless, the board did at least provide a basis for evaluating its performance with respect to price regulation by stipulating the three "tests" which it explicitly attempted, after 1967, to apply to the pricing provisions of proposed exports. These tests were originally given precise and formal definition in the course of the board's consideration of an application for export by Westcoast Transmission in 1967, although each had been applied at one time or another to particular decisions taken since 1960. By August 1970, however, the board had apparently decided to apply these three tests generally to all the applications before it at the time. It reported in its 1970 export decision:

Section 83(b) of the National Energy Board Act requires that the Board must satisfy itself that the price to be charged by an applicant for gas to be exported by him is just and reasonable in relation to the public interest.

While it must have regard to all considerations that appear to it to be relevant, the Board considers it appropriate under existing circumstances to apply to the cases now before it, *mutatis mutandis*, the criteria set forth in its 1967 decision.[22]

The board then repeated its elaboration of the three tests as follows:

(1) The export price must recover its appropriate share of the costs incurred;

(2) the export price should, under normal circumstances, not be less than the price to Canadians for similar deliveries in the same area; and

(3) the export price of gas should not result in prices in the United States market area materially less than the least cost alternative for energy from indigenous sources.

The board's application of these tests to previous export applications had varied considerably, and had evinced a desire on the part of the board to allow exceptions more often than to apply a rule. Thus, a detailed evaluation of the board's application of these criteria with respect to price is not amenable to generalized treatment, and is better recounted in the context of a discussion of the history of the four individual export complexes over the entire period under review here.[23] For present purposes, however, it should be sufficient to review the two instances noted above where the board expressly neglected to insist on the third test and approved prices below the level necessary to receive the full value of the gas in the market it served. These cases, in combination with a general overview of the board's performance, will provide some insight into the priorities that guided the NEB between 1960 and 1971. We look first at Alberta and Southern in 1970, and then at Westcoast Transmission in 1967–68.[24]

Alberta and Southern and Alberta Natural Gas

Both of these companies are Canadian subsidiaries of the Pacific Gas and Electric Company of San Francisco (PG and E), which distributes gas in California. Alberta and Southern is wholly owned by PG and E; Alberta Natural Gas is controlled by Pacific Gas Transmission Company, which is in turn controlled by PG and E. Alberta and Southern purchases gas from various producers, and then has the gas transported to the Alberta-British Columbia boundary by Alberta Gas Trunk Line. The gas is transported by Alberta Natural Gas from there to the international boundary at Kingsgate, B.C., and sold to Pacific Gas Transmission, as authorized by licences issued by the National Energy Board. From there it is carried by Pacific Gas Transmission to California where it is finally sold to PG and E. With respect to this aspect of its operations, Alberta and Southern is clearly a Canadian component of a vertically integrated, foreign-controlled corporation, as is Alberta Natural Gas.

Three striking features of the Alberta and Southern project were, first, that it was responsible for the largest single export of gas from Canada; second, that this export was made on essentially a cost-of-service basis; third, the Alberta and Southern project served only American markets. The Alberta and Southern project was permitted by Canadian authorities to export a large portion (30 to 40 percent) of Canadian exports of natural gas and, for that matter, a large portion (15 to 20 percent) of Canadian production of natural gas, at prices which for several years were below their full market value. Moreover, because of the fact that the company was engaged almost exclusively in exporting gas, the exports it made did not permit the sharing of costs between export service and Canadian service as in the case of Westcoast Transmission and Trans-Canada.

In August 1970, the board determined that the price at which Alberta and Southern proposed to sell gas would result in prices in the California market that were significantly lower than the least-cost alternative for energy in that market.[25] Since the board recognized that the application had met the other two tests with respect to price—full recovery of costs incurred and favourable comparison with prices to Canadian customers in the same area in which the export would take place—its decision regarding price was to turn on satisfaction of the third test.

While the board was willing to concede the difficulty in being precise about the failure of the Alberta and Southern application to meet this test, it did not hesitate either to declare this failure or to spell out its consequence:

> From its analysis [The Board] considers that there is in the present circumstances some gap, even though it cannot be readily identified, between the cost of Canadian gas delivered under present contractual arrangements and the cost of the lowest cost alternative energy from indigenous sources. This gap, or cost of cost of service, represents a subsidy by Canada to the United States consumers of the gas.[26]

Given this assessment, it is interesting to note the grounds on which the board ultimately ruled in favour of the application, since in doing so it was necessarily appealing to principles and criteria that overrode the three tests it had previously established for the acceptability of pricing provisions of a proposed export.[27] First, the board stated that it accepted that there is considerable force in the logic by which the applicant sought to demonstrate that the Board's "third test" . . . is not usefully applicable to an export based on cost of service, but that the justness and reasonableness of the export price in such a case must be adjudged in the light of evolution of prices over the history and foreseeable future of the project.

Second, the board pointed out that there was no easy solution to the price deficiency it had identified, since any attempt by the board to insist on a higher price at the border would have implications

so intricate that the Board has abandoned the idea on grounds of practicality and some doubt as to equity; would possibly disrupt sales arrangements already made with producers and defer producers' realization on their investment in deliverable gas; and, finally, would leave the needs of the California market for gas unsatisfied at a time when gas supply was a very grave concern for that area; . . . these results would obviously be unpalatable to all concerned.

Third,

the Alberta-California project, in which Alberta and Southern is the exporting entity, has made a very significant contribution to the development of the natural gas industry in Western Canada, and in the course of building up its highly efficient and wholly reputable enterprise has hitherto paid more for Canadian gas than it need have paid for gas from United States sources.

Fourth,

The Board is of the view that it would be inconsistent with the amity and comity which has come to characterize relations between the United States and Canada in respect of trade in natural gas to withhold approval of the last 18 to 20 per cent of the optimum through-put of the transmission system because of doubt that the price for the last increment will reflect its full opportunity cost in the California market.

Fifth, "no intervenor opposed the application and many supported it."

None of these factors was cited in terms which suggest that they were considerations which the board felt to be of general validity or applicability. Rather, they were presented in the report very much as particular circumstances which justified the neglect of general principles previously established. These principles, it may be noted, represented refinements on the board's notion of the Canadian public interest. The board seemed to be aware that in approving the price provisions of this application it had stretched the limits of acceptability under its own stipulated definition of the public interest; for it proceeded immediately from approval of this application to the warning that Alberta and

Southern and its affiliated companies should "examine alternative pricing methods with a view to establishing one more readily reconcilable with the public interest of Canada."[28]

Westcoast Transmission Company

Westcoast transmits gas out of northeastern British Columbia and a small area in the Peace River district of Alberta for export sales to the United States at Huntington, British Columbia, and for domestic sales in British Columbia. The exported gas is sold to El Paso Natural Gas Company for resale to distributors in the Pacific Northwest of the United States.[29] The Westcoast project, like that of Trans-Canada, had the dubious distinction of becoming implicated in both the political confusion surrounding the pipeline debate of 1956 and the sweeping investigations of the Borden Commission that were instigated in 1957 by a new Conservative administration.

Westcoast was incorporated in 1949 with the objective of moving gas from northeastern British Columbia and northwestern Alberta into Vancouver, interior British Columbia, and the states of Washington, Oregon, and Idaho. It appeared that the demand in British Columbia alone was insufficient to secure the company economies of scale and hence a competitive position with other fuels in its prospective markets. Exports were seen to be essential to the enterprise. However, the United States Federal Power Commission rejected the imports to the United States initially proposed by Westcoast and, when approval was finally obtained from the FPC, the border price of the gas was lower than the price charged by the company to Canadian consumers. The Canadian government, again prior to the creation of the NEB, approved this export in the hope that later gas exports would improve the situation for Canadian consumers. The board's experience with Westcoast, however, indicated that this improvement was not to be made easily.

In 1966, Westcoast instigated the first attempt to rectify the adverse terms of its gas sale to El Paso at Huntington, which had been authorized by a licence issued in 1955, by applying to the NEB for a new export licence based on a new agreement with El Paso. The board found the export price of this new agreement acceptable, and issued an export licence contingent upon El Paso receiving FPC approval to import gas under the terms of the new contract. However, the FPC in fact denied the El Paso application containing the terms of the new agreement. In particular, the FPC denied any rewriting of the terms of the 1954 agreement. Westcoast then filed an application with the Canadian board later in 1967 for, in effect, authorization to export gas on the basis of the terms set out in the FPC decision.

It was now the board's turn to find the terms of the FPC decision unacceptable.[30] The board's objection was that the price would permit

Canadian gas to enter the American market at a cost less than gas available from indigenous sources.

> The difference, some 4.5 cents/Mcf., appears to the Board to be material. It is some 14 per cent of the proposed export price, more than $8,100 per day under the Amendatory Agreement operating at a 90 percent load factor, or for the total amount of gas for which a licence is sought, over $68 million.

The board, however, conceded that "some special circumstance may justify a departure from the principle that the export price of gas should fairly reflect its value in the market area to be served." It then stated that no exception should be made in this case. The principal reason for this judgment was that the FPC had, in Opinion 526, ruled out both periodic price escalations and price renegotiations during the term of the contract. A further concern was that the price provisions suggested by the FPC in Opinion 526, which formed the basis of the application before the board, were claimed by the FPC to be "in line with comparable Canadian sales." The board was very loath to accept these conditions, partly because of fears that increasing costs of service would not be recovered by export price escalations, but also because the board did not wish to accept the principle of "in-line" pricing in relation to gas exports.

The board said that the principle that the price of exports from Canada should be "in line" with the price of comparable sales in Canada denies in principle that the price of Canadian gas should bear a reasonable relationship to the value of the gas in the market served: "The in-line principle makes the Canadian floor price the United States ceiling price." A consequence of this would be that a Canadian exporter could justify increasing the export price only by increasing his prices to Canadian customers. The board evidently feared that approval of the application would be to accept, by implication, the "in-line" pricing theory as an appropriate test by which the United States would judge the border price of Canadian imports. The board issued an order to dismiss the application.

Shortly after this decision, the NEB received yet another application from Westcoast with yet another price provision which, in effect, split the difference between the prices contained in its last two applications:

> El Paso and Westcoast in their most recent agreement, mindful of the findings of the FPC and this Board, have negotiated a price approximately midway between the prices previously put forward by the companies but rejected by one or the other of the respective national agencies.[31]

By the board's own reckoning on the preceding application (quoted above), this proposal must have entailed an undervaluation of some $35 million over the life of the contract, or of some 7 percent of the actual border price. In spite of this, and in spite of objections from Trans-Canada and B.C. Hydro, the board decided that "in the circumstances of this case, the price bears a reasonable relationship to the least cost alternative for the Pacific Northwest for energy from indigenous sources." The board went on to say that

> even if there remained a doubt that the third test as to border price had been met, the Board would be inclined to consider that, in the circumstances which have evolved, such doubt should be overweighted by general considerations of the public interest in bringing about a constructive end to a difficult matter.
>
> It is strengthened in arriving at its opinion by the views expressed by the Federal Power Commission in its order of 16 January 1968.[32]

The board then justified this easing of its hard stand on prices in terms of, first, its desire to respond positively to "a constructive initiative toward reconciliation" taken earlier by the FPC and, second, its desire to continue the board's own previous cooperative conduct and principles of conciliation for which the board had been commended by the FPC. In this connection, the board recalled a statement that it had made in an earlier decision and quoted the words it had used on that occasion in justification of the decision under review here:

> The furthest thing from the Board's mind in reaching this decision is to cause any hardship to users of gas in the Pacific Northwest. Any such effect would be regrettable on grounds both of international comity and of Canadian interests in continuing to participate in the development of the Pacific Northwestern gas market.[33]

The board's decision of February 1968 was offered as a further application of these principles.

Westcoast came to the board in 1970 with still another proposal which, in the customary fashion of Westcoast, included provisions for new exports combined with provisions for the extension of old exports, all with a bewildering variety of possible price combinations. The board found that the prices under this proposed "combined licence" met all three of its tests as far as conditions as the present time were concerned.[34] However, in accordance with objections raised by Inland Natural Gas Co., the board held serious reservations about the fact

> that if all existing export licences were to be consolidated as contemplated . . . Westcoast's export revenue, which constitutes a large part of its total revenue, would be predetermined and not subject to review over the next 19 years. Any deficiency in the return to which the Company might be entitled would have to be made up by Canadian customers since, if the [application]

were unconditionally approved, Westcoast would be free to ask the Board to increase rates to Canadian customers, but unable to alter the export price.[35]

The board's answer to this objection was to approve the application on the condition that its price provisions be amended such that "those prices would never result in an export price in Canadian currency less than 105 percent of the comparable price to Canadian customers in the area of British Columbia contiguous to the point of export of gas to El Paso."[36] It should be noted that this did nothing to ensure that the price so established would represent the value of the gas in the *American* market served, which could easily be well in excess of 105 percent of the Canadian price. The board, despite its earlier protests, had finally accepted the "in-line" method (plus 5 percent) that it had already condemned as a method that could not guarantee that Canadian gas received its full value in Canadian markets.

Pipeline Construction

All pipelines under the NEB's jurisdiction, whether they are to service export or Canadian markets or both, must receive from the board a certificate declaring that "the Board is satisfied that the line is and will be required by the present and future public convenience and necessity."[37] The board's discretion in determining what factors are to be taken into account in granting or refusing such a certificate in any individual case is practically unlimited:

> The Board shall take into account all such matters as to it appear to be relevant, and without limiting the generality of the foregoing, the Board may have regard to the following: . . . (e) any public interest that in the Board's opinion may be affected by the granting or refusing of the application.

The other considerations listed include the availability of supply, the existence of actual or potential markets, the economic feasibility of the pipeline, the financial responsibility and financial structure of the applicant, the methods of financing the line, and the extent to which Canadians will have an opportunity of participating in the financing, engineering, and construction of the line.

During the period reviewed in this chapter, the board had no opportunity either to approve or reject directly any project aimed to expand the use of oil or natural gas in Canada by extending pipeline systems farther east than existing systems already reached. However, the board did rule on several pipeline projects between 1960 and 1971 that, in effect, represented a choice between greater interdependence with the United States in energy matters and a more exclusive, strictly national approach to energy development. Two of these decisions—

Matador Pipeline and Great Lakes Pipeline—reveal the extent to which "continentalism," or the explicit commitment to cooperation and joint action with the United States in the energy field, dominated the board's thinking during this period, as some of their decisions regarding gas exports revealed. The other decision regarding a pipeline proposal to be reviewed here—Consolidated—seems to represent a departure from this line of approach, and it is included for discussion in this chapter because the NEB addressed several points in deciding this case which bear some interesting parallels to the Alaska Highway pipeline project and the "pre-build" proposal the board approved in 1980.

Matador

An early and major application of the National Energy Board Act with respect to the construction of pipelines occurred in 1960. This decision, which related to an application for the construction of an oil pipeline, led the board to define more precisely some of the responsibilities under section 44 of the act, and involved considerations that could subsequently be applied to any pipeline proposal. The application by the Matador Pipeline Company was for the construction of the facilities necessary to transport oil (delivered to the applicant at the international border by an affiliated company in North Dakota) from the border to the Interprovincial Pipeline in Canada. The Matador project was intended to be part of a system designed to carry North Dakota oil into American markets already served by Alberta crude oil, via Canada's Interprovincial pipeline.

Several considerations arose out of this proposal. First, the board was aware that

the effects of successful completion and operation of the project must include some increase in the competitive strength of North Dakota oil in relation to Canadian oil in the markets both reach, either in quantity or in price or in both.[38]

Any possible objection on these grounds was removed from the board's mind on the basis that no Canadian producer, shipper, or refiner of oil raised objections during the proceedings, and the province of Alberta submitted a letter stating it would make no objection.[39]

The second consideration was raised by an intervention in the proceedings by the Soo Line Railroad Company, which was engaged in carrying oil from North Dakota to Minneapolis-St. Paul. In what the board commended as a "skillful and learned argument," Soo contended that the Matador application showed no evidence of being "required by Canadian public convenience and necessity," thus raising a question as to the meaning of those words as they appear in section 44 of the National Energy Board Act.[40] In part, the Soo Line contended that "the question which Parliament has delegated to this Board is, does the Canadian economy require this pipe line as a pipe line in connection with the transportation needs and necessities of this country?" They submitted further that "it is not this Board's function to decide whether or not an application for a pipeline should be granted by reason of its overall effect in connection with Canada's economy and international relations."

The board took the Soo argument to be "that the Board could certificate a pipeline which is to transport oil and gas for Canadian needs only." The board rejected this definition of its responsibilities as too narrow, and stated that it rather should take into account

> the possible general effects upon Canadian oil production, marketing and transportation which in its opinion might arise from denying or approving this application even though the consideration of these effects necessarily involves some examaination of circumstances beyond the borders of Canada.

In particular, the board noted the Canadian government's declaration of the National Oil Policy and its desire to expand export sales which meant, in effect, gaining access to the United States market and in turn maintaining an exemption from American import regulations. The board stated that it must consider whether rejection of the Matador application would not threaten the success of the National Oil Policy to the extent that its success depended upon American cooperation. In conclusion, the board stated as follows:

> Rejection of this application would ill accord with the attitude which has been taken by responsible authorities in the United States to the construction of Canadian-owned pipe line facilities in the United States, the carriage through the United States of Canadian oil in Bond, and the access of Canadian oil to markets in the United States.[41]

This, in combination with the lack of Canadian objections noted earlier, seems to have been decisive for the Board. The application was approved.

Among other precedents it established, the Matador case stands as an early instance of Canadian authorities lending assistance to the United States in serving American markets with American fuels. The notion that, in so doing, they would improve rather than diminish the market prospects of Canadian fuels in the same U.S. markets seems contradictory, but it is a notion that was invoked again at an early stage in the Mackenzie Valley pipeline proposal: Canadian access to U.S. markets ultimately comes down to American goodwill, and at times this goodwill must be achieved by helping the U.S. serve the same markets with its own supplies.

Great Lakes

The principles of cooperation and the economics of interdependence in petroleum marketing were raised again in a different way—indeed, in an obverse way—in 1966 by the Great Lakes project. This proposal was reviewed by the board in the form of an application to export and re-import approximately 6 Tcf of gas and to export 765 Bcf over twenty-five years. The point at issue was whether Trans-Canada would be permitted to transport natural gas from Alberta to central Canada via the United States.

As the board itself was clearly aware, the character of this project was such as to call into question policies established during the course of earlier government decisions with respect to the original Trans-Canada system.[42] The construction of gas pipelines through the United States has always appeared to have economic advantages in terms of costs, construction, time, financing, and other factors, but disadvantages with respect to the loss by Canadian authorities of unlimited and exclusive jurisdiction over the entire transmission system. The NEB expressed reservations concerning the Great Lakes project, particularly with respect to future expansion of the Trans-Canada-Great Lakes system, which the board felt would require a degree of cooperation between authorities in the two countries that might not be entirely realistic. Specifically, approval had to rest on the assumption

> that the regulatory agencies of both countries will always move in the same direction and within a short time of one another in dealing with interrelated applications by Trans-Canada in Canada and by Great Lakes in the United States . . . This is a very large assumption and not to be taken for granted.[43]

Nevertheless, after deliberating on these and numerous other factors, the board apparently felt that these drawbacks to the scheme were overridden by its economic advantages, and were mitigated by both a sufficient degree of confidence in the continuing goodwill of American authorities and the mutual benefits to be derived from the necessary integration of the North American petroleum economy.

> The "amity and comity" in relations between the two countries in respect of gas, to which reference has been made in Federal Power Commission decisions, is real and highly valued by this Board. For its part, the Board . . . believes the growing interdependence of the two countries in terms of energy is mutually beneficial and can be made more so if its development is carefully reconciled at each stage with the national interests of the participating countries.[44]

The application was approved by the board. Following an initial period of reluctance by the Cabinet and some hurried negotiations with the industry, the applicants, and the board, the Cabinet also approved the project with the provision that a fixed percentage of Trans-Canada's total throughput to eastern Canada be carried via the northern Ontario route. Initially, this was set at 50 percent, and was eventually to reach and remain at 65 percent.

Taken together, the Matador and Great Lakes decisions reveal some of the principles and assumptions on which the board bases its judgments with respect to the construction of pipelines. These criteria are within the powers of the board to consider, but they are additional to those spelled out specifically in the National Energy Board Act. They may be summarized as follows. First, no pipeline, whether to facilitate gas exports or otherwise, should be denied on the basis that it does not appear to be essential or indispensable to Canada, but only if it is demonstrably not "desirable from the standpoint of the public interest." Moreover, in determining the desirability of the line, the board ought to consider all the effects of a denial, even though the consideration of these effects necessarily involves some examination of circumstances beyond the borders of Canada. Second, the principles of "amity and comity" between the United States and Canada, ought to govern so far as possible the relations on energy matters between the two countries, as should the general recognition that their growing interdependence in energy matters is mutually beneficial.

Consolidated

A more recent application led the NEB to still other considerations which it felt necessary to take into account in deciding on an application for a new export venture. In its report of August 1970, the board reviewed its deliberation of applications from Consolidated Pipe Lines Company and Consolidated Natural Gas Limited to construct new pipeline facilities for the export of gas from Canada, and for the construction of additional facilities to transmit gas from a gas field in Montana to their main Canadian line for eventual sale in the United States.

The volume of Canadian gas involved in the application totalled approximately 1.5 Tcf over the life of the proposed export. Because this was an application for gas exports through an entirely new pipeline in Canada, a close review of the board's considerations and findings with respect to the Consolidated project seems necessary and worthwhile from the standpoint of this study. However, the potential significance of the application in providing a clear indication of the board's applied principles and criteria is diminished by the situation existing when the application was reviewed—namely, an exportable surplus in Canada insufficient to meet the total demand constituted by all the export applications before the board. Therefore, as this discussion will reveal, it is difficult to know whether the board ultimately based its rejection of this plan on a shortage in gas supply or on a failure to meet the board's export price requirements. Nevertheless, a large number of interested parties intervened in the Consolidated application, and the case provides a picture of the configuration of interests on the gas export issue. For this reason, the present discussion will focus primarily on the board's review of Consolidated's price provisions and on the submissions made in favour of and against the project.

It had been relatively rare for large numbers of interested parties to submit positions with respect either to the export applications of particular companies or to any matters considered by the board other than its determination of an exportable surplus of natural gas. Judging from the submissions received by the board on the question of the Consolidated application, the major point of contention among the parties in this case had to do with the desirability of the entry of a large new export venture under the condition of an anticipated shortage of exportable volumes of gas. The issue, therefore, reveals the positions advocated by various parties on the questions of gas exports in a context other than that of the determination of surplus. (It also anticipates some of the conflicting interests and principles that were later to be involved in the pre-build portion of the Alaska Highway; if the entire Alaska project does not proceed, the eastern leg of the pre-built portions of the project will bear a striking resemblance to the proposed Consolidated project.)

In the majority of cases, the interested parties related their stance on Consolidated to their attitude on the proper role of the NEB with respect to export prices. It was generally accepted as a given that approval of the Consolidated project and the concomitant expansion of gas exports would lead directly to greater competition among transmission companies for gas supplies, and thereby to higher wellhead prices. Those parties who supported the Consolidated application welcomed this prospect, while those who opposed it did not. This divergence of views corresponded with another on the question as to whether the board ought to concern itself at all with the potential effects of gas exports on the price of gas to Canadians, and how the board ought to ensure just and reasonable prices for exported gas.

A clear illustration of how tightly interwoven these various questions were can be seen in the submission of the Canadian Petroleum Association. The association insisted that the board must continue to recognize as a

> very important consideration . . . the encouragement of the greatest possible degree of competition in the buying and selling of gas at the well-head which, we submit, will not only be the best assurance that the export price for surplus gas at the border will be just and reasonable in the public interest, but in addition that it will be defensible to foreign consumers since it will max-

imize the returns to producers and thereby serve as an incentive to the discovery and development of new sources of supply, a matter of considerable concern to those consumers.[45]

The CPA based its support for the Consolidated application on this principle. The association resisted any more active intervention on the part of the board in export matters:

> It would be most unfortunate if the Board implemented restrictive regulatory policies such as those advocated by certain of the Intervenors which might well be interpreted by potential investors in Canada's gas and associated industries as a trend toward economic nationalism.[46]

The substance of the CPA's position was urged on the board by all the producers who addressed the issue.[47] Further support was forthcoming from the government of Alberta, which stated simply that it supported all applicants and opposed anyone who opposed any applicant.[48] On the question of the board's intervention in the conditions of allowable exports, the Alberta spokesman confessed that he was tempted to sum up with three words: "leave us alone." Alberta took the position that arm's-length bargaining was the surest guarantee of acceptable prices, and perhaps the only one, since to obtain an opportunity price was seen to be impossible. Finally, "there is no authority in this Board to concern itself with gas prices in Canada."

A few of the parties did separate the question of Consolidated's application from the broader questions of the conditions that should be attached to exports. For example, the Saskatchewan Power Corporation was in favour of the Consolidated export, provided several conditions could be met. These included provisions that the board regularly review and amend as necessary the price provisions in export contracts, exercise a concern that the export of gas did not adversely affect the location of industry in Canada, and take steps to ensure that the return obtained from the exports would remain in and benefit Canada.[49]

Conversely, Alberta and Southern, while it did not oppose Consolidated outright, suggested a principle which under prevailing conditions was likely to lead, and apparently did lead, to the rejection of the project, namely that preference among competitors for a given volume of exportable gas should go to those systems already in existence and serving export markets. On other questions, however, the company was in broad agreement with the positions taken by the CPA, and expressly rejected the suggestion by others that there be periodic reexamination by the board of the provisions of export contracts.[50]

Numerous parties were in direct opposition to the Consolidated proposal. These parties tended to stress the implications of the scheme for the level of Canadian prices, which were predicted to rise inevitably as a result of the increased competition for gas supplies which Consolidated would bring about. Two major Canadian distributors of natural gas in Ontario, Union Gas and Consumers' Gas, added to this consideration of prices the concern that, since Consolidated would be serving the same export market as Trans-Canada, it would threaten the savings accruing to Canadian customers of Trans-Canada as a result of that company's sharing costs between its Canadian and export service.[51] On the same basis, several parties (Trans-Canada, Westcoast, Northern and Central, the province of Manitoba, the province of Ontario, and the province of Quebec) submitted that the NEB should give priority to transmission companies serving both Canadian and export markets in allocating exportable surpluses of gas among applicants, and all were opposed to the Consolidated proposal.[52]

All of these parties were joined by B.C. Hydro, Inland, and Gaz Metropolitain in supporting more active consideration by the board of the prices at which gas was being exported, particularly the effect of exports on Canadian prices.[53] More specifically, several emphasized that the board should insist on gas exporters obtaining opportunity prices in the American market they served. Still others advocated that floor prices be attached to cost of service contracts, and that periodic review of export contracts be undertaken.[54]

In the light of these conflicting arguments, the board's rejection of the Consolidated application could be taken as evidence against the case that NEB decisions taken between 1960 and 1971 reveal an export orientation on the part of the board, since Consolidated had been supported by those parties interested in expanded exports and opposed by those interested in greater protection for Canadian requirements. However, the board did not reject the Consolidated application as part of a policy to reduce the amount of gas to be exported from Canada—as already discussed, the amount of gas available for export was determined before the acceptability of individual export proposals was considered—but rather ranked the Consolidated application lowest among the various export proposals competing for a share of an exportable surplus that was not large enough to permit the approval of all the applications made. Notwithstanding some serious misgivings it had with regard to the export price provisions of the proposed project, it is not at all clear that the board would have rejected the Consolidated application if the exportable surplus had been sufficient to support all of the applications before it. According to the board's expressed reasons for rejecting the Consolidated proposal, priority must be given to pipelines serving both Canadian and export markets over pipelines serving export markets only, and to applicants proposing to export through existing facilities over those proposing to export through new facilities. In the words of the board:

Where a choice has to be made between licencing exports by a project wholly oriented to export and a project which serves Canadian customers, if all other factors were equal the choice would have to be in favour of the project serving Canadian as well as export customers; . . . When a choice has to be made between licencing additional quantities of gas for an existing project which has not hitherto been developed to optimum capacity, and licencing a new project of otherwise comparable merit, the Board considers that it is consistent with the avoidance of misallocation of resources . . . to make the choice in favour of the existing system, assuming of course that its proposal meets all relevant requirements.[55]

However, these principles clearly did not rule out for all time the construction of new, export-only pipeline systems in circumstances where a surplus of gas existed in Alberta. The pre-build portions of the Alaska Highway, approved in 1980 under just such a condition, are new facilities devoted exclusively to export service. The eastern leg of that project, in the absence of any firm and reliable link to a fully constructed pipeline to Alaska, could fairly be labelled "Consolidated II."

The Performance of the NEB to 1971: A Review

The decisions taken by the NEB and its use of criteria in deciding the acceptability of various natural gas export proposals between 1960 and 1971 support the contention that the board was primarily concerned to develop and maintain American markets for Canadian gas.[56] Taken as a whole, the record reveals that, once it had determined on other grounds that an exportable surplus existed, the board did not feel very strongly constrained by other provisions of the act, particularly those respecting the export price.

Prior to 1970, the board decided on export and pipeline applications under conditions of a relative abundance of natural gas. That is, the board had been unable to find that an exportable surplus existed which exceeded the quantities of gas for which export applications had been made to it, according to the prevailing procedures for the determination of surplus. When evaluating the individual merits of export applications under these conditions, the board's decisions showed little consistency. It is true that several principles, elaborated above, were consistently enunciated by the NEB with respect to the relevant border price provisions of individual applications. In practice, however, the board insisted upon only the full recovery by an exporter of its cost of service, and even here the special circumstances of the Westcoast project throughout the period under review seriously undermined even this "necessary" condition.

The most neglected of the three tests of the border price, however, was the opportunity price provision; that is, the test requiring that the Canadian gas realize something close to its full value in the market served. Not only did this hold true throughout the period reviewed here, but in its August 1970 decision the board allocated the full 5.8 Tcf of exportable gas available without a single cubic foot being absolutely required to meet this provision. Indeed, in the case of Alberta and Southern, the board eschewed this test consciously and deliberately. The other two major exports appeared to meet the third test under current conditions; but the future prices of the exports were tied, not to possible price rises in the American market served, but to price rises in the Canadian markets adjacent to the point of export.

The record of NEB decisions during the period from 1960 to 1971, then, shows that the only condition on which the board insisted for the export of gas was the existence of an exportable surplus. Moreover, the board permitted a gradual expansion of the proportion of total available reserves which could be exported by occasionally extending the degree to which future Canadian requirements were allowed to depend upon trend gas. In its decisions at the close of this period, the board insisted only on a price for exports defined by cost of service in two instances, and by the relation to Canadian market prices in two other instances. The board was apparently unconcerned with the overall effect of exports on either the price of gas to Canadian consumers or on the size of the Canadian market, and did not insist on gas receiving its maximum value in export markets. Where there was availability of an exportable surplus, the board revealed an inclination to allow considerations such as broadly defined cooperation with the United States or concern with the growth of the industry to override a concern with obtaining the full value of Canadian gas in American markets. Concern with, or confidence in, cooperative relations with the United States has also been a consideration in the board's decisions with respect to pipeline construction, as both the Matador and Great Lakes decisions showed.

Apart from revealing the basic orientation of the NEB, its decisions did, over the decade of the 1960s, contribute significantly to a situation that, in broad terms, was marked by these features, among others:

1. The continuing export of large volumes of lower-cost readily accessible gas reserves and a corresponding reliance on higher-cost, relatively inaccessible gas reserves for the protection of future Canadian requirements;

2. The possibility that a natural gas pipeline from the Arctic would be required to ensure for Canadians incremental supplies of natural gas, predictably at substantially higher prices;

3. The pricing of exports on the basis of Canadian prices, so that the export price depended, not upon the value of gas in the export

market but rather upon prices—wellhead prices in two cases, and market prices in two other cases—in Canada;

4. A necessary reliance on American goodwill for the dependability of one imporant pipeline link between Canadian sources of natural gas and the principal Canadian market for natural gas (Great Lakes).

These decisions and conditions were significant departures from what was held to be in the Canadian public interest by the Borden Commission, a number of academic observers, some members of Parliament, and a variety of other groups and individuals who had spoken publicly on these issues. Moreover, on several occasions, the board explicitly overrode principles it had itself defined to be in the public interest; and it did this with reference to other principles which it never formally stated to be of general applicability. A search through the NEB reports of this period for the overriding principles in reference to which it justified these decisions in favour of exports will turn up most frequently the following.

1. The principles of "amity and comity," or mutual trust and goodwill, between American and Canadian authorities and interested publics on matters pertaining to the development, production and marketing of natural gas;

2. The principle that gas exports to the United States can assist Canada in providing service to Canadian markets more cheaply than would otherwise be the case, largely through the medium of economies of scale on pipeline facilities;

3. The principle that exports of natural gas encourage the development of a resource and of regions of Canada which otherwise could not be economically developed;

4. The principle that the fact of past service and the historical terms of service ought to be considered in relation to incremental service to the same market.

The Performance of the NEB: Parliament's Reaction

Since the four principles set out above defined, to all intents and purposes, the outcome of Canadian energy policy of the 1960s, it is important to review briefly the reaction (or rather lack of reaction) of Canada's parliamentary representatives to the NEB's exercise of its responsibilities. While it is true, as we have seen above and as many have said in recent years, that the National Energy Board decided energy issues during that decade almost exclusively in response to the representatives of the private firms and provincial governments that took part in its proceedings, it is also true that there was no significant degree of public pressure on the board or on the government in general to adopt a different course. With one notable exception—which should have a very familiar ring to it—Parliament virtually ignored the NEB and its exercise of its authority over the disposition of Canada's natural gas resources. The exception was the Great Lakes project, Trans-Canada's proposal to expand the capacity of its system by constructing a pipeline through the United States to serve export markets and the growing market in Ontario.

The case of Great Lakes Transmission prompted what was without doubt the most prolonged and heated parliamentary debate on a board decision to take place in the 1960s. There is evidence that the Cabinet reversed its original decision to reject the project, which had been approved with reservations by the board, after extensive meetings with representatives of industry.[57] A compromise plan was worked out, according to which the southern line would be allowed but would, according to a letter of agreement between Trans-Canada and the government of Canada, always carry less than one-half of the company's total deliveries to its markets in eastern Canada.[58]

The government stated its rejection of the original proposal on August 25, 1966, and announced its approval of a modified proposal on October 4. The debate on the government's new stand began in earnest the day it tabled copies of the formal agreement between the government of Canada and Trans-Canada Pipe Lines Limited. The debate was originally in the form of comment on the ministerial statement. From October 31 to mid-November, the debate continued on the vehicle of an interim supply bill for the Department of Defence.

John Diefenbaker, as leader of the opposition, stated his immediate reaction to the tabled documents:

> We find today that continentalism has become the keystone of this government's policy. Continentalism was preached in the days of Macdonald. He stood against it and assured us that Canada would be an independent nation.[59]

T. C. Douglas of the NDP pledged that

> if the government proceeds to put this agreement into effect and to authorize Trans-Canada Pipe Lines to build this line through the United States, I maintain it is doing so in contravention of the Trans-Canada Pipe Lines Act, that it has no power to authorize the construction of this line without submitting the agreement and the letters of intent to parliament for approval, and . . . we shall insist that this matter be submitted to parliament.[60]

Later in the debate, another member of the NDP insisted that "this government does not have a mandate to create a continental economy."[61]

Highly reminiscent of the pipeline debate of a decade earlier, Great

Lakes revived the question of whether eastern Canadian markets should be served by means of a pipeline controlled entirely or only partially within Canada. The main arguments in favour of the southern route were lower construction costs and reduced transmission costs owing to the possibility of combining exports with Canadian service in the same project.[62] The main arguments against the route were the loss of exclusive Canadian government control over the flows of gas through the Trans-Canada system and the loss of the potential stimulus of an expanded northern line to the economy of northern Ontario.[63] Objections were occasionally raised that the export component of the southern route, contrary to its claimed benefits, would give industries in the Great Lakes states an advantage over their Ontario competitors.

Of the two other types of decisions the board must take—export prices and the size of the exportable surplus—export prices have provided more grist for the parliamentary mill than has the determination of surplus. But the opposition in Parliament to the actions of the NEB, and to their subsequent approval by the government, was feeble by comparison with the strong resistance mounted against some of the pipeline proposals debated in the House during the early 1950s. A detailed review of the reaction in Parliament to each of the decisions taken between 1960 and 1971 reveals that, in keeping with what might be considered a tradition, members of Parliament were primarily concerned with pipelines, and in particular the routing and ownership of pipelines. They devoted less attention to the other questions addressed in this chapter. With the exception of periodic complaints about the export prices charged by Westcoast Transmission (which might also qualify as a minor parliamentary tradition), the export decisions of the NEB and the policies of governments during the period under review received relatively scant attention.[64]

The conclusion seems warranted, then, that the bulk of the membership in the House of Commons—that is, the members of both the government and the opposition—sustained for over a decade attitudes on the issues of pipelines and natural gas exports that were satisfied in a general way by NEB decisions. If anything, at the close of the 1960s and the beginning of the 1970s, the Liberal government and the minister of energy, J. J. Greene, were even more enthusiastic about the prospects of a continental energy program than the board appeared to be in its reports.[65] Meanwhile, the failure of the official opposition to take the government to task for its approval of board decisions is consistent (with one exception) and striking. Perhaps the most notable example was opposition leader Robert Stanfield's failure to even mention the board's August 1970 decision—the largest single volume of gas exports ever approved by Canadian authorities—in his address to the speech from the throne, which he made within days of the decision.[66] (T. C.

Douglas of the NDP had moved the adjournment of the House to discuss the same decision at the close of the previous session a few days earlier, and had been overruled by the Speaker on the grounds that an opportunity to debate the issue would be available at the opening of the next session.[67]) Other examples may be found in the occasional combined votes of the government and the official opposition against motions by the CCF (and later the NDP) in these matters, which go back as far as the bill to establish the National Energy Board.

To observe this situation of near bipartisanship is one thing; to account for it is another. An important factor, presumably, is the fact that Cabinet decisions to approve NEB actions do not come automatically before Parliament, and therefore require extra effort to bring before the House for debate. Another suggestion is that potential opposition to the actions of the NEB had been lulled to indifference by the "symbolic reassurance" provided by the independence and expertise of the board in exercising its regulatory functions.[68] An example of this would be the CCF's acceptance of the first exports approved by the board, which was justified on the basis of the independence and technical competence of the board and the fact that numerous interested parties who had been represented before the board had not opposed the exports.[69] Finally, it might be supposed that the two major parties have simply been influenced in one way or another by the industry. This could well be true, but it is important to note in this connection that organized and vocal opposition to board decisions was not very frequent or powerful *outside* Parliament until the 1970s.

Whatever may be the strength of these and other possible factors, the truth remains that the board ran into little trouble from Parliament, either in the form of opposition moves to embarrass the government into rejecting the board's recommendations, or even in the form of casting, during the period under review, the proceedings and decisions of the board under the light of public scrutiny. Some aspects of this situation were to change drastically during the course of the following decade. Certainly a larger number and a greater diversity of groups were to descend upon the NEB in an attempt to influence its decisions and reports. The amount of time devoted in Parliament to issues bearing on the transportation and marketing of fuel resources in Canada was to increase exponentially. But despite this enormous quantitative increase in the public's attention to these matters in the 1970s, on at least one issue the nonpartisan quality of parliamentary reaction to board decisions and related matters persisted. The board's approval of the Alaska Highway pipeline in preference to the original Mackenzie Valley pipeline received support in principle from all parties in the House of Commons as a means to carry Alaskan gas to American markets.

7 National Self-sufficiency and North American Interdependence: Canadian Energy Policy in the 1970s

LARGELY as a consequence of the series of increases in the world price of crude oil during the 1970s, and particularly of the Arab oil embargo of the fall of 1973, debates on issues relating to oil and gas in Canada increased in number and intensity as the years went by. Some of these issues, such as the setting of a uniform price below the world price for oil across Canada, have no direct precedent in the years reviewed earlier in this study. Other issues, such as the extension of the interprovincial oil pipeline to Montreal and the proposed extension of the Trans-Canada system for natural gas into Atlantic Canada, are directly related to themes and policies already discussed: the less reliable or more expensive foreign supplies of fuel become, the greater the inclination of the federal government to intervene to expand the market in Canada for Canadian fuel. Yet, the most striking fact about Canadian energy policy during the 1970s was the resilience of some of the traditional constraints bearing on the creation of a national fuel market, despite the fact that international developments had established that foreign oil supplies were neither more reliable nor less costly than Canada's domestic fuel supplies.

The Arab oil embargo of 1973 and the ensuing escalation of the world price of oil resulted in growing public support for the general objective of Canadian self-sufficiency in fuels, a goal dramatically characterized in one oil company advertising campaign as "the big, tough, expensive job of reducing Canada's dependence on foreign oil." As always, however, the realization of self-sufficiency in fuels in Canada (or, what is the same thing, the creation of a nationwide market for Canadian fuel sources) necessitated major initiatives with respect to the transportation of fuels, and the government's commitment to national self-sufficiency partly translated into a commitment to three pipeline projects aimed to expand the Canadian consumption of Canadian fuels: the extension of the interprovincial oil pipeline from Sarnia to Montreal to displace about 250,000 barrels of foreign oil per day out of Canadian markets east of the Ottawa valley; the creation of a northern pipeline to facilitate Canadian access to its own Arctic supplies of natural gas by means of a joint Canadian-United States project designed primarily to carry Alaskan gas to the lower forty-eight states; and the extension of the Trans-Canada system to permit the substitution of Canadian natural gas for foreign oil in most major centres east of Montreal. Nevertheless, by 1981, roughly eight years after the government announced a comprehensive new energy strategy aimed at reducing Canada's dependence on foreign oil, the geographic extent of the Canadian market served from domestic fuel sources had increased only marginally: only the Sarnia-Montreal oil pipeline had become a functioning reality. The reasons for this failure during the 1970s to provide Canadians in every part of the country with reliable access to Canadian fuels, despite conditions in the international market that were incessantly described as a "crisis," will be explored here in a review of the politics surrounding the northern pipeline, the Sarnia-to-Montreal pipeline, the Petroleum Administration Act, the pre-building of the Alaska Highway pipeline, and the National Energy Program of 1980.

The Mackenzie Valley and Alaska Highway Pipeline Debates

The federal government's promotion of the idea of a northern pipeline combining deliveries of Alaskan natural gas to the American markets with deliveries of Mackenzie Delta-Beaufort Sea gas to Canadian or American markets, has a long history—so long, in fact, that its justification of the project has had an opportunity to undergo three different phases. It was first endorsed by the Canadian government as a means of maintaining markets in the United States for Canadian oil and gas, and of obviating the necessity of tanker traffic down the British Columbia coast. Then, as oil and gas shortages loomed in Canada, it became the most efficient means of gaining access to new supplies of natural gas for Canadian markets. Finally, as the Canadian component of the project declined in both likelihood and significance, it became a means of providing a substantial stimulus to the Canadian economy.[1]

As early as 1971, for example, Jean Chretien, Minister of Indian Affairs and Northern Development, informed a group of petroleum engineers in Dallas, Texas, that it was clear to him that

> if the oil and gas reserves of Prudhoe Bay are to be brought to market, they will have to come part, if not the whole way by pipeline. In the case of natural gas, I think that it will have to be moved all the way by pipeline and that the likely market would be in the mid-continent region. We in Canada welcome the building of such a pipeline through our country and would do anything reasonable to facilitate this particular development.[2]

These were the days in which the Canadian government was offering the Mackenzie Valley to the United States as a corridor for the transmission by pipeline of both oil and natural gas. The offer of an oil pipeline had already been made as an answer to the concern that Canadian oil would be shut out of some American markets if Canada did not cooperate in the task of delivering Alaskan oil and gas to the lower forty-eight states, and to the added concern that the British Columbia coast would be threatened by oil spills if Alaskan oil was transported by tanker to the Pacific states instead of by pipeline across Canada.[3]

Two years later, in a major speech on energy policy made in the wake of the Arab oil embargo, Prime Minister Trudeau placed before the House "proposals which will set the basis for a new national oil policy." It seems fair to say that the flagship in this newly launched fleet of initiatives was the construction of an oil pipeline to Montreal, a policy that would "abolish the Ottawa Valley Line." However, Trudeau also noted the importance of "encouraging the substitution of relatively more plentiful resources for relatively scarcer ones," and proceeded to point to the "enormous quantities of gas . . . available to be transported from the far north."[4] He continued:

> A major development is the proposed gas pipeline up the Mackenzie Valley to move Alaskan gas to United States markets and at the same time to make it possible to move Canadian northern gas to Canadian markets . . . The government believes that it would be in the public interest to facilitate early construction by any means which do not require the lowering of environmental standards or the neglect of Indian rights or interests.

The overall objective of this policy, and the goal to which all the other proposals were directed, was simply "Canadian self-sufficiency in oil and oil products," an objective to be reached before the end of the decade. The other proposals included development of frontier oil and gas, a national petroleum company to assist in this development, support for the development of the Athabaska tar sands, and a commitment to higher oil and gas prices. Implicit in this breakdown of the government's approach was the idea that self-sufficiency in fuels was to be achieved in part through the substitution of natural gas—and to a large degree, Mackenzie Delta–Beaufort gas—for oil in Canadian markets served at the time by imported oil.

There can be no doubt that the most controversial of the new initiatives outlined by Trudeau in December 1973 was the proposed northern pipeline. Between 1973 and 1978, the Mackenzie Valley version of this project (more precisely, the Canadian Arctic Gas proposals or CAGPL) was the subject of countless questions in the House of Commons; an enormous set of studies conducted by the sponsoring consortium, government departments, independent academics, and the sponsors of rival projects; an impressively comprehensive program of hearings conducted by a major inquiry; an equally impressive and exhaustive report by that inquiry; an unusually long series of hearings and an extraordinarily comprehensive report by the National Energy Board; and even a few demonstrations and public meetings sponsored by a variety of public interest groups.[5]

However, despite the almost unprecedented amount of attention given to the Mackenzie Valley pipeline question, compared with all earlier Canadian pipeline decisions, including Trans-Canada, the government's ultimate approval in 1978 of the northern pipeline in the form of the Alaska Highway project (this time sponsored by Foothills Pipe Lines) was received by Parliament and the Canadian public alike almost without opposition.[6] One reason for this was that the opposition to the Mackenzie Valley line had rested on objections to particular characteristics of the Mackenzie project rather than to the essential nature of the northern pipeline concept. Hence, emphasis had been almost exclusively placed on the damage that the construction of a Mackenzie Valley pipeline might cause to the native communities, animal life, and natural environment in the western portions of the Canadian Arctic. Considerably less criticism had been levelled at features of the Mackenzie Valley project that would necessarily be involved in any project designed to carry Alaskan gas across Canada to American markets, such as the desirability or otherwise of the project's macroeconomic impact on Canada, or the considerable increase in Canadian-American interdependence that the project seemed certain to bring about. Moreover, little of consequence had been made of the contention that the Canadian stake in the project (from the standpoint of providing Canada with access to new fuel supplies) was highly questionable, given the relatively small volume of proven reserves of Canadian gas in the western Arctic that could be delivered through the proposed system to Canadian markets, and given the probability that rising gas prices in Canada and the United States would substantially increase supplies available in Alberta.[7] In short, the opposition to the project argued that it was a bad thing because it was a hazard to native people and wildlife in the Northwest Territories, rather than because it was unnecessary or detrimental to the Canadian national interest in energy terms or in broader economic terms.[8] This line of argument on the part of opponents of the project left the promoters of a northern pipeline with two possible ways through (or around) their objections: with enough care, attention, planning, and regulation (read the Northern Pipeline Agency) the pipeline along the originally proposed route could be made a "safe" pipeline; or, the pipeline could be built along another "safer" route. When Justice Berger's report on CAGPL made a politically compelling case that the first of these choices was impossible, the government turned to the second.[9] The Alaska Highway project ultimately became the political "end run" around the opposition to Mackenzie, and thereafter all the government had to worry about was whatever opposition might remain to a "safe" northern pipeline, that is, a pipeline that would perform the same function the Mackenzie Valley line had been designed to perform, but with relatively fewer of its potentially harmful effects on the human communities and natural environment along its route. There was little such opposition.

Realistically or not, the northern pipeline was no longer seen as a danger to either people or places along its route, and only three kinds of objections could now be raised in opposition: concerns relating to its macroeconomic impact on the Canadian economy; doubts about its ability to contribute anything commensurate with its cost to Canada's capacity to meet its own energy needs; and concerns about its implications for Canada's independence of the United States in energy matters. These concerns had also been raised in certain quarters in relation to Mackenzie, but they had been completely overshadowed by other concerns localized in the Arctic region. Ironically, by the time the northern pipeline issue reached Parliament in the form of a bill that could be subjected to formal debate, the macroeconomic impact of the line was perceived as one of its major virtues, if not its only one from a purely Canadian standpoint; its impact on Canada's energy position was largely immaterial, partially because there was a good chance the line in its new manifestation might never carry any Canadian gas; and the project was almost universally endorsed as a worthwhile, friendly gesture to the United States. (In its new form, the northern pipeline was to proceed across Alaska, the Yukon, British Columbia and Alberta, via a route that comes no closer than 500 miles to Canada's gas reserves in the Mackenzie area. The "Dempster Spur" was supposed to link these reserves to the main line, thus graphically illustrating what a northern pipeline has always represented: an American project with a Canadian appendage—even if the Dempster Spur is ever built.) About the only way the northern pipeline could have been taken any farther from its earlier ostensible contribution to energy self-sufficiency for Canada would have been to use it to justify the export of natural gas from Alberta to the United States. This wasn't done until 1980.

Meanwhile, in 1978, the northern pipeline controversy entered a new phase in the form of a debate over Bill C-25 to establish a Northern Pipeline Agency to oversee the construction of the Alaska Highway project. It is worth noting that the government's objective of self-reliance had recently been given more precise definition—the government had set an import target of one-third of Canadian consumption or 800,000 barrels per day by 1985—and the role of a northern pipeline in meeting this objective through the displacement of foreign oil with Mackenzie gas had been given more explicit and emphatic recognition. In support of the pipeline bill, the minister of energy, mines and resources summed this matter up, as follows:

Canada still does not have enough gas to meet its long term needs, where "need" is here defined so as to include a rapidly increasing market share for natural gas. The current rates of discovery in Alberta are certainly encouraging, and it would indeed be fortunate if large, new accumulations proved up in Canada's western basin. But I do not think we can afford to take such a gamble. We cannot afford to say that we are sure that these reserves will be proved up. It is quite clear that we need to develop other gas producing areas in Canada as well, and we have to take advantage of every opportunity we have in this country. The Mackenzie Delta-Beaufort Sea region is one such area of large new potential, and Canadians simply cannot afford to ignore it. Access to the gas in the Mackenzie-Beaufort is crucial to Canada's energy future. It is for this reason that I urge an early passage of the bill now before us.[10]

The speech had earlier established the points that "our dependence on oil supplies from the Middle East must be reduced" and that "natural gas is one of our great opportunities to reduce dependence on imported oil."

The interesting point about this argument in favour of the Alaska Highway project is not that it was made once again by the minister of energy, but that it was ignored by almost everybody else, including other ministers. Indeed, in moving the motion, Deputy Prime Minister Allan MacEachen spent relatively little time on the relationship between the northern pipeline and Canada's energy needs compared with that spent on elaborating various terms and conditions that would ensure Canadian participation in the construction of the line and protect native interests and the environment. Given his speech to the motion, it would appear that the primary virtues of the government's action on the pipeline were, first, that "the launching of this mammoth northern pipeline project will provide a much needed boost to consumer confidence and business confidence in both countries, one that will go beyond the very considerable real impact it will have in stimulating production and employment directly and indirectly" and, second, that "the conclusion and implementation of the bilateral agreement [Canada's pipeline agreement with the United States] are important also as a means of further strengthening the bonds which have traditionally linked Canada and the United States closely together."[11]

At any rate, it was absolutely clear from the debate to follow that the potential of the project to provide Ontario (and other Canadian) manufacturers with orders, and Canadian workers with jobs, assumed much greater significance in the eyes of almost the entire membership of the House than did its potential to provide Canadian consumers with fuel. This shift in emphasis in evaluating the Canadian interest in the project—a shift from a concern about energy supply to a concern about economic stimulus—is sharpest in the case of the NDP. T. C. Douglas, the party's principal energy critic at that time, expressed the view that the project was to all intents and purposes an exclusively American venture and could not be even remotely justified as serving Canada's energy interests. This was because, first, there was not enough Canadian gas up there to matter very much and, second and even more important, what Canadian gas could be carried by the pipeline was not needed in Canada. However, Douglas argued that, for these very reasons, the government must attend closely to the terms under which the pipeline was built: if all Canadians were getting from this project is employment and dollars, he seemed to be saying, at least get all the employment and dollars that were there to be got. In his own words,

> It should be remembered that the primary purpose of this pipeline is to take American gas from Alaska to the United States, and if no Dempster spur is built Canadians will never get a cubic foot of that gas. This is a neighbourly gesture to the United States. Therefore, we have taken the position that, if we are going to provide our American neighbours with a land bridge by which they can take gas from Alaska to an energy-hungry nation, then Canada has a certain right to benefits from that project . . . What happened to those benefits, Mr. Speaker?[12]

The remainder of his speech can be fairly described as an indictment of the government for failing to extract terms from the American government that would provide the full potential economic benefit to Canada arising out of the construction of the project. In the event, the NDP, while supporting the concept of the Alaska line, voted against the bill on the ground that it did not provide sufficient protection for Canada's economic (as distinct from energy), social, and environmental interests.[13]

The immediate response of the Progressive Conservatives to MacEachen's motion was to censure the NDP for what they feared would be a filibuster of the bill by that party.[14] While expressing reservations concerning some aspects of the bill, the Conservatives were basically concerned that the project face no extended delays. In the words of Erik Nielsen, the first Conservative member to address the bill, "this bill can do more to accomplish a sense of purpose and unity in this country than anything that has transpired during ten years of Trudeau musings and of a tired and worn out government."[15] The only amendment they supported on division in the House was one of their own to the effect that Ontario ought to be included among the provinces represented on the federal-provincial consultative council which was to be struck under the provisions of Bill C-25.[16] They voted against several NDP amendments that went to division, and they voted for the bill on final reading.[17]

Conservative support for the notion of a northern pipeline, unlike the position of the NDP, apparently derived from an interest in both the economic stimulus to be provided by the project and its potential contribution to the expansion of the Canadian petroleum industry in the western Arctic. Harvie Andre (Calgary Centre), for example, expressed his view that the pipeline was a necessary but not a sufficient means of meeting Canada's future gas requirements and recommended further measures, such as relaxing land-use regulations pertaining to northern lands under federal jurisdiction, and allowing accelerated exports from Alberta. He asked the House to remember that "with a swap arrangement or an acceleration of existing exports we do not all diminish our long term security position. We simply provide cash flow at this point to those who are demonstrably using it to find more gas."[18] Meanwhile, Perrin Beatty of Ontario was lauding the pipeline not only because it promised adequate energy supplies in the future, but because "it will also have industrial spin-offs in all parts of Canada and will have enormous benefits in terms of business in the country, trade, and badly needed jobs."[19] It is hard to say for certain which of these two major interests in the construction of the Alaska pipeline counted as the decisive one in the eyes of the official opposition, but some indication may be found in the fact that on the same day on which they voted to support the bill, one of their members was expressing some doubt that the Dempster Spur would be built.[20] Apparently, even if the project promised no guarantee of access to Canadian energy supplies for Canadian use, it was worthy of Conservative support.

The Sarnia-Montreal Pipeline and the Petroleum Administration Bill

The other major initiatives undertaken by the government in response to the oil embargo and price hike of 1973 resemble the northern pipeline project, at least in so far as they were aimed to protect Canadian consumers from the potential harm they might suffer as a result of the unreliability or high prices of foreign crude oil. The Montreal pipeline would clearly do this, simply by supplying part of the Canadian market east of Cornwall, Ontario, with Canadian rather than overseas oil, its traditional source of supply.[21] The Petroleum Administration Bill (C-32) allowed the government to maintain a uniform price for oil across Canada, meaning that Canadians who continued to depend on imported oil would not pay higher prices than Canadians who had access to cheaper domestic supplies. It also empowered the federal government to set a price for oil across Canada that, although lower than the prevailing world price, would nevertheless be sufficiently above pre-1973 levels to encourage the development of new Canadian supplies.

The Montreal Pipeline

The government's decision to promote the extension of the interprovincial pipeline from Sarnia to Montreal proved to be the least contentious provision of the government's 1973 energy program. This could have stemmed from the fact that the decision was formally the responsibility of the NEB rather than Parliament, so that no bill on the matter was ever introduced to the House, but it is also the case that the principle of extending the market in Canada for Alberta oil as far as Montreal received the approval of both the NDP and the PCs. If anything, in fact, the government came in for mild criticism for what opposition members saw as unjustified reluctance on the part of the government to grant the assurances or subsidies which the Alberta producers in general, and Interprovincial Pipe Lines in particular, demanded as the price for their cooperation in the venture.[22] The industry wanted to know either that the Montreal market would be guaranteed for a period of time sufficient to write off the new pipeline or that it would be designed to be reversible in the event that for reasons of price or suppy the Ontario market was given over at some future point to imported oil. They also demanded a subsidy on the operation of the line.[23] Finally, in keeping with an earlier phase in the construction of Canadian pipelines, there were calls from both sides of the House for a decision to supply Montreal with Canadian crude oil via an all-Canadian route.[24]

The Montreal pipeline, however, bore an even closer relationship to themes discussed earlier in this study. Once operating at full capacity, it would create a situation in which, for the first time, both the Ontario and Quebec markets would be served with substantial Canadian supplies of their primary fuel source. (It is true that, as far as simply providing both the Montreal and Toronto-centred fuel markets with physical *access* to Canadian fuel supplies is concerned, the Sarnia-Montreal pipeline achieved for Canada with respect to oil what Trans-Canada had achieved almost two decades before with respect to natural gas. However, the market for natural gas east of the Ottawa Valley was comparatively small throughout the 1960s and 1970s.) In this connection, it is interesting to note that, as earlier with natural gas, the Liberal government's adoption of this solution was not its first preference. Rather, its inclination was to explore a continental swap with the United States. According to Canada's energy minister, who was providing the background to the government's decision in principle to support the Montreal line,

the obvious and first starting point was to know whether an agreement could be achieved with the government of the United States in terms of assuring supply to eastern Canada, either from United States sources or from sources controlled by the United States in the event of interruption. The obvious quid pro quo in this situation was that we have a very substantial delivery of oil into the United States market which at the present time is 1.2 million barrels a day, and the undertaking would be to assure that supply to the United States if they could supply a comparable assurance to us in Eastern Canada. In fact, it was not possible to get an American assurance in that regard.[25]

The Petroleum Administration Bill

The debate on the federal government's decision to seek the power to regulate the prices at which Canadian oil and natural gas are sold in Canada concerned a number of issues for which the earlier debates on Canadian fuel policy provided little or no precedent. In particular, some Progressive Conservative members—frequently but by no means exclusively from Alberta—sustained a direct attack on Bill C-32 for its alleged unconstitutionality, which was to say its alleged violation of the principle of provincial control of natural resources.[26] However, the debate on the bill also involved two long-standing issues in Canada's fuel debates: national unity and national self-sufficiency. While addressing these familiar themes in relation to national fuel policies, the speakers for and against the proposed Petroleum Administration Act did so in a manner which reflected a fundamental change in the circumstances surrounding Canada's fuel problem—a reversal of the traditional relationship between the price of imported and domestic sources of fuel. The national-unity issue no longer had to do with how far central Canada would go in conceding a share of the Canadian market to domestic producers, but rather had to do with how far producing provinces would (or should) go in supplying those markets at prices below their market value. Thus, from the vantage point of the producing provinces, national unity and Canadian fuel policy had less to do with the plea to "provide us with a stake in Confederation in the form of a market for our fuel supplies" than with a protest: "What has the rest of Canada done, or what is it prepared to do, in recognition of its stake in our fuels?" Or, in terms of the actual debate, what was to be the compensation to fuel-producing provinces for Canadian sales of domestic oil and gas at prices below world levels?

Meanwhile, on the self-sufficiency issue, the age-old trade-off between the reliability of Canada's fuel supply and its price no longer related to a division of the Canadian market between domestic and foreign sources of supply, but was instead related to the *price* at which domestic fuels would supply as much of the domestic market as was practicable. As always, the consumer interest in the lowest possible price was in conflict with the consumer interest in assured, long-term supply, but now in an entirely different way. If, as signified by its commitment to supply both the Ontario and Quebec markets with domestic oil, the government no longer sought a balance between the cost and

reliability of the country's oil supply by drawing a geographic demarca-
tion line between areas of Canada to be supplied by cheaper foreign oil
and more reliable domestic oil, it would attempt to strike a similar
balance by establishing a price-level for oil across Canada at a point
somewhere between the old Canadian price and the new, higher world
price. The Macdonald price replaced the Borden line.

The politics of the Petroleum Administration Bill at any rate
represented the same conflict between producer and consumer interests
over the proper level for the Canadian price that had occurred before
over the extent of the Canadian market for the Canadian fuels. Firms,
governments, and MPs from the producing provinces tended to argue
the merits of a higher price in ensuring long-term security of supply,
while their counterparts from the consuming provinces—especially the
central Canadian ones—argued the merits of a lower price in holding
down the cost of living and maintaining the competitiveness of Canadian manufacturing industries. As usual, the government struck—or
claimed to be striking—a compromise between these two considerations, and located the "national" interest in a trade-off between maximum reliability and minimum price; that is to say, a price "high
enough, but no higher than the level required to bring forward an adequate supply for Canadian requirements."[27] According to the federal
minister of energy, the established price at that time ($6.50 per barrel)
not only reflected a consensus among the first ministers of Canada as to
the appropriate price to obtain until July 1, 1975, but also reflected "a
reasonable balance between the interests of consumers and producers in
Canada."[28] However, lest this consensus break down at some point, and
lest it should prove difficult to maintain a price "determined at mutually
acceptable levels by agreement with [the producing] provinces," it was
deemed prudent that the federal government, as the final arbiter of provincial interests, "should equip itself with powers to provide for oil
price restraint . . ."

The official opposition objected to the bill on precisely these
grounds; they did not like to see the federal government "taking unto
itself the authority to set the price without reference to anybody else
. . ." As the minister of energy correctly pointed out, "the essence of
the difference" between the Liberal government and the Conservative
opposition concerning the bill was the provision that final authority
over the setting of oil and gas prices in Canada reside with the federal
government, a provision Robert Stanfield saw as part of "a two-pronged
massive assault on provincial control and development of natural
resources."[29] The minister neatly underscored this difference:

> The leader of the Opposition argues that prices should be set by negotiation.
> But what if negotiation fails and a producing province insists on going to

world prices? Is a single province, then, to exercise a veto over the Canadian markets?[30]

One need not endorse the substance of the government's position on this score to give it credit for at least stating its position on the appropriate balance between consumer- and producer-province interests more forthrightly and less ambiguously than various spokesmen for the official opposition. Both the repeated calls for negotiation and consultations coming from the Conservative members and the conflicting (or at best, extremely vague) positions they took with respect to the desirable price level for oil in Canada create the impression that the party was either unable to resolve the conflict between consumer-province and producer-province interests within its own ranks, or (probably more likely) was attempting to promote, or at least defend, producer interests without appearing too obviously to be abandoning the interests of consumers, especially those in Ontario.

In defending their view of the consumer interest, Conservatives tended to place greater emphasis on the national interest in long-term security of supply than did the Liberals, who usually spoke more of the consumer interest in restraining price increases. Sinclair Stevens, for example, expressed the warning that

> without provincial control over price-fixing mechanisms, the producing provinces will be unable adequately to ensure a fair rate of return to the provinces and oil companies which are developing petroleum resources. Without adequate return there will be a marked decrease in exploration activities and consequent decrease in supply to the consuming provinces.[31]

To the undiscerning reader, this statement might appear to be a criticism of the federal government for seeking control over oil prices with the intent of keeping them too low to bring on new reserves in sufficient volume to assure domestic supplies for Canadian consumers. When an equally undiscerning listerner, Donald Macdonald, charged Stevens with having just made a ringing appeal for higher prices for the consumers of Ontario, Stevens rose on a point of order to deny he had ever suggested higher oil prices.

Some Conservative members, particularly those from Alberta, opposed the government's plan to control prices at all, and openly called for rising prices in the interests of Canadian self-sufficiency. According to Stan Schumacher of Calgary, for example,

> surely it is in the interest of this country that we be self-sufficient and not be at the mercy of other nations. It would be better to pay a higher price and have an assured supply than to take the short run benefit of a lower price and not have a supply.[32]

The NDP expressed less dissatisfaction with the principle of the bill than had the Conservatives during the debate. T. C. Douglas, the party's energy critic, stated early in the debate that the NDP not only endorsed both the policy of a single Canadian price for oil and the policy of keeping that policy below world levels, but had been recommending such a course of action for some time.[33] Moreover, the NDP gave its explicit support to the idea that the federal government should have the legislative power to set oil prices across Canada, "as that is the only way we could have a two-price system in Canada."[34] The only reservations the NDP seemed to have about this issue evidently stemmed from the fear that the federal government, in exercising its power over prices, might allow the Canadian price to rise beyond the point justified by the need to generate funds for future exploration and development, and called for some assurance that all extra money generated by future oil price increases would be controlled by either the federal government or the producing provinces "rather than leaving it to the good intentions and tender mercies of the oil industry." The NDP, in short, sided with the government much more than with the official opposition and, if anything, made fewer nods in the direction of producer interests than the federal government appeared ready to do. Their position was clearly based on the belief that long-term self-sufficiency was realizable at much lower prices than either of the two major parties would concede.

So much for the self-sufficiency argument and the Petroleum Administration Bill; the implications of the bill for national unity or, at least, for equity in the terms of confederation, were also debated, albeit in a rather one-sided fashion. The use of the term "one-sided" may seem unjustified in view of the fact that clear differences existed among members of Parliament as to the fairness of the two-price system: should the oil-producing provinces be expected to subsidize Canadian consumption of oil by accepting a reduced rate of return on both Canadian and export sales of their oil? Despite conflicting views on this question, it is interesting that almost without exception, increases of any magnitude in the Canadian oil price—whether they were to go all the way to the world price or were to be held well short of that level—tended to be supported solely on the basis of their contribution to Canada's future security of supply: higher prices were reluctantly or enthusiastically endorsed, but in either case were endorsed only to the extent that they would contribute to the *national* interest in the long-term availability of supplies. There were few voices expressing the idea that higher prices were justified simply by virtue of making their owners, the citizens of Alberta and Saskatchewan, rich.

Nevertheless, the arguments of a few members came close to such a proposition. These arguments held that, if the producing provinces were, for the sake of central Canadian consumers, to remain less pros-

perous than they could become by charging the going world rate for
their depleting natural resources, then central Canada should be
prepared to contribute to the prosperity of the producing provinces in
other ways, such as by reducing the protection central Canadian industries enjoyed in the form of tariff structures and national transportation policies that descriminated against the West. As Doug Roche put it
at one point in the debate,

> what is at stake here is the legitimate regional and provincial development of
> our country. I laugh when I hear speeches about Alberta and its strong energy
> position and the suggestion that it is adopting a separatism [sic] position
> going its own way. Nothing could be further from the truth. Far from trying
> to get out of Canada, members from Alberta for years have been trying to get
> into Canada. We have been trying to assert ourselves as real partners.[35]

Later in the same speech, he went on to say that the minister of energy

> has tried to make it appear as if we are opposed to the interests of central
> Canada. I am not opposed to the interests of central Canada, but I am opposed
> to our entire country being dominated by a philosophy which protects cen-
> tral Canada at the expense of the development of jobs and resources in
> western Canada as well as in the maritimes. That is the issue.

The response of the government and, indeed, of central Canada to this
line of argument was basically that the present direction of federal
policy was simply a matter of tit for tat: when Albertans complain that
they cannot sell oil to Canadians at world prices, the minister

> would point out that there has not been a tradition of selling oil to Canadians
> at world prices. For ten or fifteen years oil was sold to central Canada at
> about $1.25 to $1.50 above world prices.

Moreover,

> The Alberta industry developed on the basis of a series of federal policies, not
> central Canada policies, but federal policies which substantially assisted the
> development of that industry and from which at that particular time the pro-
> ducers of Alberta benefited.
>   That was due to the fact that Ontario in particular instead of taking the
> offshore oil which could have landed more cheaply into the Toronto and Sar-
> nia refineries, developed industry on the basis of western Canadian oil. There
> was no question of victimizing anybody. It was a good policy, and it assisted
> the development of the Alberta industry.[36]

On this point, the minister had already received the endorsement of Max
Saltsman of the NDP, who estimated the total cost to Canadians of
eleven years of the National Oil Policy to be between $500 million and
$5.5 billion.[37]

In the end, the Petroleum Administration Bill was passed with the support of the two opposition parties. It could be concluded, therefore, that the interests of central Canada prevailed in this debate, in that it obtained the power to acquire access to Alberta's oil on federal terms, if need be. However, it could also be said that Alberta's interests were at least partly respected, in that Canadian prices were subsequently allowed to move toward international levels, although disputes on that score remained troublesome, to say the least, for the remainder of the decade.

However, the remarkable thing about Canadian energy policy during the remaining years of the 1970s and the beginning of the 1980s was not this failure of the federal government and the producing provinces to agree amicably on the proper price level for Canadian oil and gas or on the appropriate distribution of revenues from prevailing prices; the conflicting interests and perspectives that surfaced in the course of the debate on the Petroleum Administration Act could be expected to diverge even more starkly along similar lines as international prices continued to escalate and the economic stake of producing and consuming provinces in these matters became even higher. The remarkable thing was, rather, the government's failure to bring the country any closer than the Sarnia-to-Montreal pipeline brought it to a state of genuine self-sufficiency in fuels. With the arrival of the 1980s, in fact, Canada was still importing roughly 400,000 barrels per day of foreign crude oil into markets east of Montreal; the future of the Trans-Quebec and Maritimes pipeline, designed to assist the substitution of Canadian natural gas for some of that imported oil, was clouded in uncertainty; new gas exports to the United States were approved for delivery by means of new pipelines optimistically described as the "pre-build" portions of the Alaska Highway system; and a new national energy program was announced that, despite the apparent "nationalist" inspiration of some of its provisions, portended yet another phase in the development of what has been termed here as a "quasi-national" or "semi-continental" pattern of fuel transportation and marketing.

The Pre-build Decision

A central issue in Canadian energy policy during the 1970s culminated in the government's decision in the summer of 1980 to approve an amendment to the Northern Pipeline Act permitting the "pre-building" of the southern sections of the Alaska Highway Pipeline from gas fields in Alberta and British Columbia to American markets, a decision that included a commitment of Canadian gas exports to be delivered through the new pre-built facilities as a means of improving the financial viability of the overall project.[38] This decision gave rise to some con-

troversy in Canada as to whether it was more likely to expedite the completion of the larger project, or to bring about its possible cancellation
or delay by reducing the urgency of the American need for Alaskan
gas.[39] While it was obviously impossible to predict at the time whether
or not the Alaska pipeline would actually be built, it did appear that the
Canadian government's approval of the pre-build project in the absence
of absolute assurances that the larger project would be undertaken constituted a significant shift in government priorities from a desire to
achieve the economic stimulus promised by the construction of the entire northern pipeline to a desire to improve Canada's balance of energy
trade and to dispose of a growing surplus of gas in Alberta by means of
expanded exports. In this respect, the pre-build decision can also be
viewed as yet another example of the way in which Canadian energy
policy has traditionally been caught between the economic advantages
of international trade in fuels and the political advantages of interprovincial trade in fuels, and how the energy needs of the United States
have affected the resolution of this tension in Canadian policies toward
the trade and transportation of Canadian fuels. At the time the government approved the construction of the pre-build project, the eastward
expansion of the Canadian market for Alberta natural gas beyond Montreal to Quebec and the Maritimes was still faced with regulatory delays
and surrounded by doubts about its economic viability.[40] In other
words, it appeared once again that a significant expansion in the extent
of the Canadian market for Canadian natural gas was more difficult to
realize and was a less immediate priority than an expansion of the export market for such gas. In this sense, the pre-build decision represented not only the culmination of a major issue in Canadian energy
policy during the 1970s, but also the latest phase in the entire history of
Canadian energy policy. To understand this, it is necessary to review the
three possible outcomes of the government's decision of July 1980, as
well as the consequences associated with the most probable of these
outcomes.

The only certain outcome of the Canadian and American approval
of the pre-build proposal was the construction of two new pipeline
systems for the transmission of new exports of Alberta natural gas to the
American midwestern and Pacific states, respectively. Decisions taken
in July 1980 appeared to assure this, but to guarantee no more than
this.[41] If, despite all the public and private assurances to the contrary,
the construction and operation of these systems did not contribute as
planned to the eventual completion of the entire Alaska Highway pipeline—if "pre-build" turned out to be simply "built"—the result would
be a new natural gas transmission system whose exclusive function
(from the Canadian perspective) was to export Canadian natural gas sup-

plies to the United States. Such a development might be termed the "export-only" venture.

A second possible consequence of the approval of the pre-build proposal, and the one most clearly intended by the corporate sponsors of the scheme as well as the governments of both Canada and the United States, was the eventual construction of a pipeline connecting the natural gas fields of Alaska with the pre-build portions of the project, via a route traversing Alaska, the Yukon Territory, the northeast corner of British Columbia, and western Alberta. If this project proceeded as expected, the result would be a transmission system whose exclusive function (save for relatively small volumes of Canadian exports during the years of the pre-build phase) was to deliver American gas to American markets. Such a development might be termed the "transport-only" venture, denoting the fact that it would represent the provision by Canada of a transportation facility (or land bridge) by means of which the United States could gain access to its own supplies of natural gas.

A third possibility was the eventual construction of a spur line or lateral connection between the main Alaska Highway pipeline and the Canadian natural gas deposits in the area of the Beaufort Sea and the Mackenzie Delta, commonly referred to as the Dempster Spur. If this component of the Alaska project was completed, the result would be a transmission system that combined two functions, the delivery of Alaskan gas to American markets and the delivery of Canadian Arctic gas to Canadian markets. Such a development might be termed the "joint-service" venture.

Of all that could be said about the differences among these three possible outcomes of the Canadian government's pre-build decision, it is sufficient here to note, first, that each scheme entailed a different set of effects on the supply position of Canada and the United States. If the entire system was in place within five to ten years, it would mean that the United States would be supplied with gas from Alberta for the time it took to complete the northern sections of the line—a relatively modest amount of Canadian gas—and thereafter would be supplied with relatively large volumes of natural gas from Alaska. In short, the impact of the prompt completion of the joint-service project on the American supply position would be to increase American imports from Canada in the near term and, in the longer term, to increase American access to its domestic supplies and possibly to reduce its dependence on Alberta gas. Its impact on the Canadian supply position would be the converse of this: exports would increase for a brief period and could decline again when the line was completed, while Canada's access to domestic supplies would be increased by the construction of the Dempster Spur.

The effects of the other two forms of development differ significantly. This is especially true for the export-only venture: neither Canada nor the United States would increase its access to gas supplied under its own jurisdiction, while American imports of Canadian gas (and, correspondingly, Canada's export commitments of Alberta gas) would be much greater, including not only the five to ten years of deliveries envisaged in the pre-build proposal but probably also a volume of exports sufficient to continue the same rate of deliveries for another ten to fifteen years at least. (If the larger project were to fail, deliveries of Canadian gas by means of the pre-built facilities would be likely to continue in order to extend the utilization of the pipelines and to avoid imposing possible hardship on the American buyers who would have become dependent on the supplies of gas being delivered through them. As we have seen in chapter 6, such an attitude toward continuing exports through existing pipelines on the part of the Canadian authorities would not be unprecedented.) The differences between the transport-only venture and the joint-service venture are not as marked as those between the joint-service venture and the export-only venture. The dropping of the Dempster Spur would make little difference to the implications of the project for the American supply picture, but the construction of the line to Alaska without the Dempster Spur would mean, of course, that Canada had gained nothing from the northern pipeline in the form of access to new Canadian gas supplies. From the standpoint of their potential impact on Canada's ability to be self-sufficient, though, the differences between the export-only venture and the transport-only venture are still quite significant, for they predictably amounted to the difference between long-term and temporary commitments of Alberta gas to American markets, and could potentially affect the future availability of Alberta gas supplies for present and potential Canadian markets. Moreover, of these two versions of the project, only the transport-only venture promised the kind of economic stimulus that had become, by 1978, the principal justification for Canada's whole participation in a northern pipeline.

It is worth recalling that when the Northern Pipelines Act was passed in 1978, the importance of the construction of the project in providing Canada with an economic stimulus was at the forefront of arguments made by all political parties to justify their endorsement of the principle that the line should be built. Some members, including the minister responsible, made passing reference to the by then largely eclipsed objective of gaining access to Canada's Arctic gas reserves, but the emphasis was almost always placed on either the jobs to be generated by the construction of the line or on the goodwill and cooperation that Canada should continue to extend to the United States in energy matters. Indeed, the NDP, the only party in the House to vote against the bill on third reading, did so precisely because it did not go far enough, in their eyes, in guaranteeing substantial returns to Canadian industries and workers out of the pipeline's construction. For the same reason, the NDP attempted to obstruct the government's approval of the pre-build scheme in July 1980 in the absence of "iron-clad" guarantees that the entire project would soon be undertaken. Without such guarantees, the NDP and other critics of the pre-build scheme insisted, the government was giving the go-ahead to an export-only venture rather than to the preliminary stage of either the transport venture or the joint-service venture, and the economic activity generated by the construction of the export-only venture was not of sufficient scale to justify the associated exports of Alberta natural gas that they thought should otherwise be retained for Canadian use.[42]

For their part, the government and other supporters of the early start on the pre-build portions of the Alaska Highway argued that sufficient assurances that the entire line would be built had already been provided by congressional resolutions and presidential letters expressing endorsement of the northern pipeline and that, apart from this, the export provisions of the pre-build scheme would provide an important means of disposing of Alberta's growing natural gas surplus. Further delays in the construction of the line, such as those that would occur if the government made its approval contingent upon final financial arrangements in support of the entire project, would have meant that a large number of gas producers would be left with substantial volumes of unsold gas, depressing the rate of exploration and development in the industry. In this way, the justification of the pre-build project had come full circle to one of the justifications of the original Mackenzie Valley proposals: the Canadian oil and gas industry must not be left to languish in a state of excess capacity.

Lost in all of this was the argument that had been tirelessly put forward by both the government and the industry in support of the Mackenzie Valley pipeline during the mid-1970s, an argument that can be summarized as follows: Canada needs to lessen, if not eliminate, its dependence on unreliable and costly overseas oil; one important step in the direction of fuel self-sufficiency is the substitution of Canadian natural gas for foreign oil in Canadian markets not now served (or not as fully served as they could be) by Canadian natural gas; one of the most promising sources of supply of natural gas for Canada lies in the western Arctic; the most efficient means of tapping this new source of supply is a northern pipeline combining deliveries of Alaskan gas to American markets with deliveries of Mackenzie gas to Canadian markets. It is perhaps not surprising that the growing surplus of natural gas in Alberta took some of the edge off the importance of the Mackenzie Delta as a new source of Canadian gas supply and, therefore, off the immediate urgency of a northern pipeline in the form of the Alaska Highway line including the Dempster Spur. It is less obvious why this growing surplus of natural gas in Alberta was not committed (as Mackenzie gas was supposed to have been committed) to Canada's goal of self-sufficiency rather than to the American market.

Given all this, it seems fair to say that Canadian approval of the pre-build project in the absence of reliable arrangements for the completion of the larger project constituted a softening of the government's commitment to the objectives of facilitating Canadian access to Arctic supplies by means of a joint-service pipeline, stimulating the Canadian economy by means of the construction of the overall project, and generally expanding the Canadian consumption of Canadian fuels. The government did this in favour of the short-term disposal of Canada's current natural gas surplus through increased exports to the United States market, and of the reduction of Canada's net deficit in energy trade through increased exports rather than through import substitution. This is clearly a re-application of precisely those principles that have traditionally combined to produce a continental rather than a fully national market for Canada's fuel supplies.

Immediately following the pre-build decision, it was not altogether clear whether U.S. decisions affecting the fate of the northern pipeline would be made soon or whether, when taken, they would favour an early start on the full Alaskan pipeline. Neither of the assurances provided by the president and Congress to the Canadian government included a commitment of financial support from the U.S. government for the construction of the line, and the private sponsors made no unconditional commitment to an early start on the project.[43] The uncertainty on the American side had to do with the future marketability of Alaskan gas at the price this gas would have to obtain if the Alaska Highway pipeline was to pay for itself. As in western Canada, recent price increases for natural gas in the United States were bringing on an increase in the supply available from its conventional producing areas, to the point where the American market for Alberta gas at prevailing export prices was softening, making the market prospects for much more costly Alaskan gas even less promising. Ironically, the Canadian exports approved as part of the pre-build scheme could, as some Canadians feared, have further reduced the urgency and apparent practicality of the larger system if they were taken as evidence that Alberta gas would be available at prices below the prospective delivered price of Alaskan gas for the foreseeable future. In sum, the market prospects for the throughput of the proposed Alaska pipeline were not bright when the Canadian government proceeded with the pre-build portions of the line, and that decision did nothing to improve them. Of course, financial guarantees for the line from the Canadian or the American governments could not be ruled out, nor could other forms of action such as a government-set price structure that might have improved the apparent viability of the project. Nevertheless, while it may indeed have amounted to a "judgment call" to predict whether a decision in favour of the pre-build project would do more on balance to improve or diminish the probability that the Alaskan line would be built in the near future, there could be no doubt that the decision made the expansion of the export market for Alberta natural gas more immediate and more definite than it did the expansion of the eastern Canadian market for the gas.[44]

The Canadian government went ahead with the project in spite of these uncertainties, and in doing so declined to adopt either of the most obvious alternatives to its action: to hold the surplus gas in Alberta until the entire Alaskan project had obtained full financial backing, or to delay the construction of the pre-build pipelines until the completion of the Trans-Quebec and Maritime project. In this sense, the most obvious beneficiaries of the decision were the gas producers of Alberta. They would move their gas more quickly and in greater volume by this action than by any other. In addition, to the extent that the federal government imposed an export tax on this gas, Canadian taxpayers would receive some help in carrying the burden on general government revenues represented by the subsidy on imported oil. Finally, all of those interests that stood to gain from the eventual construction of the Alaska line will have benefited from this decision if, after all, the pre-build exports do contribute to the successful completion of the northern pipeline: the sponsoring company in Canada (Foothills) and its co-sponsors in the United States; American distributors of natural gas and their customers; the United States as a community and the American government in its quest for self-sufficiency; Canadian and American construction companies and manufacturing firms and suppliers to those concerns; the workers in all of these industries; and all the others whose interests would be served in some manner by the largest single private investment in history. The only people conspicuous in their absence from this list were those Canadian consumers of overseas oil who were still without access to Canadian natural gas to meet part of their fuel requirements, and Canadians generally, who remained no less vulnerable to whatever catastrophe could befall the citizens of a country whose government leaves its foreign policy hostage to potentially hostile foreign suppliers and a substantial portion of its citizens dependent on a highly unstable foreign region for its fuel supply. This seems a curious oversight on the part of a government whose energy policy for almost a decade had ostensibly placed a very high if not an absolute priority on energy self-sufficiency.

Although the fact remained that measures to expand the export of Alberta natural gas to markets in the United States were approved before measures to deliver Alberta natural gas to markets in the Maritime provinces were finalized, it was not possible to argue conclusively that the pre-build exports actually threatened natural gas service to the Maritimes or would hamper Canada's capacity to reduce its dependence on overseas oil. Indeed, as if to allay any lingering doubts along such lines, the government followed the pre-build decision with a comprehensive package of new energy policies and initiatives, nearly all of which were justified with reference to their potential contribution to the realization of national self-sufficiency in fuels, and several of which were aimed to expand the use of Alberta natural gas in eastern Canadian markets as far as Halifax.

The National Energy Program, 1980

Since its release in the fall of 1980, the most widely discussed and highly controversial provisions of the government's national energy program (NEP) have been those relating to the Canadianization of the oil and gas industry; to the diversion of the exploration and development activities of that industry away from the currently producing provinces toward Canada Lands; and to a significant increase in the federal share of oil and gas revenues.[45] Nevertheless, there can be no question that the pervasive emphasis throughout the NEP was on the importance of national self-sufficiency in fuels. One indication of this was given in the first of three "precepts of federal action" listed by Energy Minister Marc Lalonde in his introduction to the program: such action "must establish the basis for Canadians to seize control of their own energy future through *security* of supply and ultimate independence from the world oil market."[46] More than a declaration of intent, the program also included several immediate and concrete steps in the direction of energy independence, including incentives to assist homeowners and businesses in converting from oil use to gas and electricity, and support for the extension of Canada's natural gas pipeline system to the Maritime provinces.[47]

For these and other reasons, the NEP was generally described as a major departure on the part of the federal government and was both hailed and (in some quarters) condemned as a step in the direction of economic nationalism, particularly with regard to its provisions for the Canadianization of the oil and gas industry.[48] Certainly from the perspective of this study, the construction of the Trans-Quebec and Maritimes gas pipeline, which would represent the creation of a nation-wide market for a domestic source of fuel for the first time in the nation's history, would have to count as a new and significant development: interprovincial trade in fuels would finally have prevailed over the combination of forces that had hitherto limited east-west trade across Canada in favour of north-south trade across the international border. Nevertheless, there were indications that the NEP would not necessarily result in the expansion of Canadian trade in fuels to the *exclusion* of continental trade in fuels, and the future implementation of the provisions announced in the program seemed likely to involve a continuation and even an intensification of the forms of Canadian-American cooperation in the energy field that have prevailed throughout the history of Canadian fuel policy, a likelihood stemming from what the NEP did and did not say about the future development of oil and gas in Canada's frontier areas—that is, the Canada Lands that the program favoured so substantially above future oil and gas development in the currently producing provinces.

The government did not rule out the future export of oil and natural gas from these potential producing areas. Having stated the principle that, at the very least, frontier supplies ought to provide Canadians with a "safety net" for the future, and having posed the question, "Can such a principle be reconciled with exports of northern oil and gas?", the program went on to say:

> Many Canadians are understandably sceptical about assertions that Arctic resources should be produced quickly for export, as if energy were a commodity like any other. If energy were an ordinary commodity, Canadian taxpayers would never have supported provision of the rich incentives that have been available to the petroleum industry. Canadians would want to be sure, in the event that any of these reserves were judged surplus to domestic needs, that broad social and economic benefits justified their sale to others. The Government of Canada recognizes these concerns and will be very demanding in its assessment of export proposals.[49]

To the uninitiated, this may have sounded reassuring, but read in conjunction with some knowledge of an earlier phase in Canadian oil and gas development, this passage at the very least left the door open for oil and gas exports to (most likely) the United States from the frontier regions, if for no other reason than the traditional one of defraying part of the enormous cost involved in transmitting these remote resources to central Canadian markets. On this score, the NEP had nothing to say directly; indeed, the program was surprisingly silent on the central problem of bringing these new sources of supply to established Canadian markets. Like many academic studies of Canadian energy policy, the report revealed a tendency to regard energy self-sufficiency as a matter of ensuring that Canadian supply equals Canadian demand, when a major part of the problem with energy in Canada is, as it has always been, a matter of ensuring that Canadian supply is *connected* to Canadian demand, usually by means of an extremely large and costly transportation system. To help reduce these costs, such systems have had to be constructed on a large scale, often larger than the Canadian market alone would warrant: hence, exports. There was no reason to assume that the future marketing of the resources developed on Canada Lands would not be subject to the same constraints.

In fact, as a supposedly nationalist document, a strategic master plan for the achievement of energy security for Canada, the NEP revealed some serious shortcomings, particularly with respect to the role frontier resources were to play in securing Canada's energy future in comparison with reserves of natural gas already developed in the currently producing provinces and already connected to major Canadian markets. If Canada's future fuel supply was in such a precarious position that huge investments in the exploration, development, and transportation of frontier resources were justified, why did Canada continue to export roughly one-third of its current gas production from the much more favourably located gas fields of Alberta? Or, put another way, if gas being exported from Alberta and British Columbia was indeed surplus to "reasonably foreseeable future Canadian requirements," why did the government see fit to tax Canadians in order to subsidize the development of frontier supplies, either directly through Crown corporations or indirectly through generous tax incentives to private firms? With this implicit refusal to consider setting aside established reserves of natural gas in Alberta and British Columbia for the expanded or prolonged use of such gas by Canadians rather than exporting them because they appeared to be surplus to Canada's currently foreseeable needs, the government was continuing a traditional policy that dates back at least as far as the Borden Commission. But, as we have seen, those were the days when the government was working from a declared commitment to continentalism, not nationalism, and to the objective of providing assured markets to Canadian producers, not providing assured supply to Canadian consumers. If the declared intentions of the government had thus taken a 180-degree turn, it seemed reasonable to wonder why there was so little change to be observed in what the government was actually doing.

From the days of the Borden Commission, the Canadian nationalist position has always been that proven, low-cost, and readily accessible reserves of natural gas in western Canada should not be exported, leaving Canadians to depend on costlier and more remote supplies for their future requirements. The worst fears of such nationalists would have been confirmed by two policies adopted almost simultaneously by the government in 1980: the approval of increased exports of Alberta natural gas via new pipelines to the United States (pre-build) and increased public subsidization of the accelerated development of still-inaccessible, high-cost fuel supplies in the frontier regions of Canada. Of course, a refusal to allow exports during the 1950s and 1960s would have had consequences that many—and, at any rate, the government—

considered undesirable: the Canadian petroleum industry would have been left in a state of excess capacity, and would have had to bear the cost of earlier investments in oil and gas development upon which they would have received no immediate return; the expansion of the Canadian industry would have been retarded and could have been expected to take place only at a rate commensurate with the comparatively slow rate of growth in Canadian demand for fuels; and Canadian consumers would have been denied the savings available to them through the construction of transmission systems serving both export and Canadian markets. In short, exports were once given priority over the prolonged conservation of the resource in order to promote the *development* of oil and natural gas production and transportation capacity at a time when the Canadian market for fuels was limited. By 1980, however, the petroleum production and transmission industries ranked among the largest in Canada; the Canadian market for natural gas had grown and was destined by virtue of the government's own policies to undergo further significant expansion in both geographic extent and share of the market; and the government had endorsed the "novel" idea—dating at least as far back as the Borden Commission hearings—of purchasing gas from (Canadian) firms in short-run difficulty over markets.[50]

In these circumstances, the notion that the Canadian petroleum industry should expand no more rapidly than the increase in demand for natural gas in Canada, and that the long-term availability of natural gas for Canadians should have priority over the short-term income from sales to export markets might have suggested itself to a government whose overriding concern in energy matters was declared to be security of supply for Canadian consumers. But the decisions and declared policies of the government in the second half of 1980 instead committed to export markets a substantial portion of the proven gas reserves to which Canadians already had access, and directed the exploration and development activities of the industry toward potential reserves to which Canadians had no access, and to which access would only be achieved, if at all, through transmission systems combining further exports with deliveries to Canadian markets. Such policy initiatives may or may not have promised Canada's "security of supply and ultimate independence from the world oil market," but they did seem certain to ensure that the transportation projects required to bring Canadian supplies to Canadian markets would continue to depend upon the policies of the American government for their timing and the terms and conditions under which they would operate. As a form of precedent for such major undertakings, the Canadian experience with the Alaska gas pipeline might have indicated the kinds of adjustments in Canadian priorities that North American cooperation in these matters could entail. But the question of whether or not a traditional goal of policies aimed at self-sufficiency in fuels—enhanced national independence—would in fact be served by this degree of North American interdependence is not one that a study of the history of Canadian fuel policy can put to rest. Nevertheless, it may be worth recalling that North American interdependence was precisely what some of the earlier Canadian calls for a national fuel policy were aimed to prevent.

8   The Politics of Energy in Canada: Some Conclusions

THE FOREGOING study has examined the extent to which successive Canadian governments since Confederation have taken measures to expand the consumption in Canada of Canadian fuels, as well as the public controversies that the adoption (or rejection) of such measures has engendered. The study provides the basis for some general observations concerning Canadian fuel policies over the years, and the different and sometimes conflicting considerations and interests that have been brought to bear on the question of meeting the fuel requirements of the country from its own sources of supply. It also provides the basis for some conclusions about the reasons for the observed patterns of policy and politics. Admittedly, however, this study is longer on description than on explanation. It shares this characteristic with many, perhaps most, Canadian studies in public policy, particularly those that can fairly be identified as "case studies." Nevertheless, the fact that the preceding chapters have examined Canadian fuel policies over time and with respect to different fuels means that they will sustain more comprehensive general observations and less speculative (though certainly only partial) explanations than one might otherwise expect. Since the value of such an approach can only be demonstrated by its results, this discussion will recapitulate briefly the patterns in Canadian energy policy described in the earlier chapters, and conclude with an analysis of the conditions and variables that appear to account for them.

Energy Policy and Politics: A Review

Developments in the world petroleum market during the 1970s made energy self-sufficiency at least the ostensible goal to which nearly all the efforts of the federal government in the energy field were to be directed. Self-sufficiency became the standard against which the desirability or otherwise of every move in the energy field was to be decided. Energy self-sufficiency became an unquestionably Good Thing, a high, if not vital, national priority, the ultimate end of public and private endeavour in the area of energy production, transportation, and marketing. A vast army of oil company executives, politicians, federal and provincial bureaucrats, editorialists, academics, and ordinary citizens began to march behind the banner of energy self-sufficiency, and only a few stragglers even asked whether or not Canada could realistically hope to achieve this goal; almost no one questioned whether it was desirable, a goal worth striving for. Since a national consensus appears to have formed around the desirability and importance of meeting Canada's fuel requirements exclusively from Canadian sources of suppy and eliminating Canada's dependence on imported oil, and since Canada's past policies in this direction have been the focus of this study, it would seem appropriate to close this discussion by underscoring a few propositions concerning the role of fuel self-sufficiency in the history of Canadian energy policy and of the country.

Three such propositions are sufficient for present purposes: (1) Canada has never had fuel self-sufficiency; (2) some Canadians in some part of the country or another have always advocated, while others have opposed, the practical measures necessary to achieve it; and (3) those who have supported such measures have tended to support them only to the extent that they could find somebody else to pay for them. In sum, far from being an absolute goal, national self-sufficiency in fuels historically has not been even a preferred goal of Canadian governments and of most Canadians most of the time. Rather, the preferred goal of the federal government and of most Canadians has usually been to achieve satisfactory and reliable trade arrangements on an international basis: producers have favoured American and other international markets; consumers have favoured American and other foreign suppliers; the government has pursued policies that have partially respected these preferences and partially neglected them in favour of some overriding national interest involving greater Canadian consumption of Canadian fuels. It should come as no surprise, then, that government policies aimed at increasing Canadian consumption of Canadian fuels have come about frequently as a response to some form of dislocation in international trading arrangements, and have never been met with universal support, much less enthusiasm, because they have seemed inescapably to benefit one section of the country at the expense of another. However, if this analysis is correct, it could be asked, What factors changed to permit a consensus in support of national self-sufficiency to emerge during the 1970s?

There are two plausible answers to this question. One answer is to stress the significance of the reversal, after 1973, of the traditional relationship between the price of imported fuels and the price of Canadian fuels. The other is to argue that the observed consensus, even in circumstances in which international oil prices are far higher than the delivered cost of Canadian fuels, is more apparent than real, and that many of the endorsements of the objective of self-sufficiency are more conditional than they might seem at first glance. Whenever Canadian fuels delivered to major Canadian markets cost more than reliable foreign fuels delivered to those same markets, it is not surprising that the only people in the country to support the substitution of Canadian for foreign fuels are those with an interest in the production of Canadian fuels, especially when they have no other or better outlet for that production. Self-sufficiency costs consumers. It should be equally predictable that when the reverse is true, and Canadian fuels can be delivered to Canadian market at a cost below that of foreign supplies, the consumer objection to the use of Canadian fuels will disappear; and, other things being equal, the producer and consumer interest will converge on the desirability of expanding as far as possible the Canadian consumption of Canadian fuels, so that the virtues of self-sufficiency or of a national fuel policy are equally extolled in the consuming and the producing provinces (assuming, of course, that adequate international markets at higher prices are not available to producers). Such was the case in the 1920s with respect to coal, and such was the case in the 1970s with respect to oil and gas. However, in neither of these instances did agreement on the importance of self-sufficiency as a goal prevent disagreement over the practical steps necessary to bring it about. (Nor, incidentally, did the consensus in either case translate into any immediate federal action toward its full implementation.) When the reliability or the price of imported fuels makes them less attractive to Canadian consumers than Canadian supplies, the regional conflicts over whether or not to adopt a national fuel policy are simply displaced to a regional conflict over the terms and conditions under which such a policy is to be put in place.

This was as evident recently as it had been in the past. After 1973, Ontario acknowledged the importance of the goal of self-sufficiency, but argued that it must be achieved at prices below world-market levels and below the level that producers and the producing provinces claimed was necessary. The province also balked at paying higher prices for natural gas delivered by Trans-Canada Pipelines in order to allow that company to reduce the costs of its prospective service to the Maritime provinces, a step which appeared to be necessary to promote the substitution of Alberta natural gas for overseas oil east of Montreal. The producers, meanwhile, cited the vital importance of national self-sufficiency to justify rapid increases in the Canadian price of oil and gas, but insisted on various subsidies and guarantees from the federal government before proceeding with the extension of the Interprovincial oil pipeline from Sarnia to Montreal. They have also pressed the federal government to commit Alberta's present natural gas surplus to export markets rather than to an expanded Canadian market. For its part, the federal government, which justifies nearly every move it makes in the energy field with reference to the contribution to be made to the achievement of self-sufficiency, appears content to watch Canada's national oil company prepare for liquefied natural gas exports and conventional gas exports to the United States, and to allow the National Energy Board to delay the construction of a natural gas pipeline to the Maritimes while approving "pre-build" exports to the United States. Countless examples such as these show that the consensus behind the goal of fuel self-sufficiency for Canada merely obscures the conflict of interests occasioned when concrete measures are taken toward its realization. It is an unvarying tradition. Canadians in their various regions or sectors have never supported self-sufficiency absolutely, but only conditionally—the condition being that someone else be found to bear the extra outlays or forgone opportunities associated with it. Extolled from time to time throughout Canada's history for its potential contribution to national unity and national independence, a national fuel market for Canada seemed no closer to realization after a tenfold increase in the world price of oil than it had been several years after the coal-scarce winters of 1921-23; and the gains to national unity that have been derived from the federal government's most recent energy programs in the service of national self-sufficiency have been, to say the least, difficult to detect. If, in these ways, national dependence and national disunity are among the most enduring characteristics of policies to promote Canadian consumption of Canadian fuels, it is perhaps not surprising that even to the present day moves in this direction have been tentative, partial, and of only marginal effect. What conclusions can be drawn from this about the forces and circumstances that have determined the size of the Canadian market for Canadian fuels?

Energy Policy and Politics: An Explanation

In an article published in 1976, Richard Simeon set out an "assessment and critique" of policy research in Canada that, among other things, argued the case for a number of improvements in the way such research is conducted.[1] In essence, Simeon argued that the central concern of policy studies should be to explain why governments do some things and not others, a task that should include providing "much better descriptions of what governments actually do" and going beyond case studies to "longitudinal studies of the evolution of policy over long periods."[2] He set out several categories or types of causes to which reference should be made in attempting to present a reasonably complete explanation of government action: the environment; the distribution of power; prevailing ideas; institutional frameworks; and the decision-making process.[3] While the present study of Canadian fuel policy can scarcely claim to measure up to all of Simeon's rather exacting standards, it does examine Canadian policies with respect to a given policy area (national policies toward the trade and transportation of Canadian fuels) over a long period of time (from Confederation to 1980) and with respect to different fuels. It thus provides the basis for stronger conclusions about the role of the more general and pervasive forces in the shaping of Canadian energy policy, such as the environment, the distribution of power, and ideas than it does about the role of more specific and immediate influences, such as institutions and processes. This study is less concerned with decisions than with *patterns* of decisions and with the relationship between policies and "the circumstances under which [they] were initially developed," the "political forces arguing for and against them," and "the justifications by their proponents" that Simeon suggests can help in forming judgments concerning what governments are really up to when they do what they do.[4] More specifically, this study has examined the circumstances under which successive Canadian governments have acted to expand the Canadian consumption of Canadian fuels. The degree of such government intervention is the dependent variable in this study; what is being explained here are the variations or differences in the degree of government intervention, for in this study the dependent variable does change, both over time and with different fuels. The object, of course, is to relate similarities and differences in the degree of government intervention to similarities and differences in conditions and circumstances prevailing at the time in a manner that supports an inference that the latter caused the former. This exercise will conclude the study.

The information presented in the preceding chapters supports the conclusion that the major determinants of Canadian policies toward the trade and transportation of Canadian fuels are, first, the geographic distribution of North American fuel deposits and major centres of fuel consumption; second, the import and export policies of the American government; third, the concentration of political power in central Canada, especially Ontario; fourth, the extent to which the political goals of national unity (particularly as this goal may be served through the promotion of regional development) and national independence (particularly as this goal may be served through the reduction of the country's reliance on foreign supplies of fuel) have been perceived to justify the economic costs inherent in interprovincial as compared with international patterns of trade in fuels. Of course, this is not to say that there are no other significant influences at work when governments adopt measures affecting Canadian consumption of Canadian fuels; but it is to say that the four factors just identified do condition demands that set the agenda for government and severely constrain the policy alternatives that are likely to appear feasible and desirable. The question of the relative significance of the various factors determining Canadian fuel policies is probably best examined by using Simeon's "funnel of causality" to set out more systematically the relationships between different types of policy determinants.

The Environment

As stated several times previously in this study, the single most important factor affecting Canadian policies with respect to the development, transportation, and marketing of energy sources in Canada is the fundamental geographic one of the distance between Canadian sources of supply and Canadian markets. But the importance of this "environmental" factor is even greater when combined with another, namely, the closer proximity of American fuel supplies to Canada's largest single fuel market and the closer proximity of large American fuel markets to Canada's fuel-producing regions. The consequences of this elemental geographic reality for Canadian energy policy have been substantial: first, international trade in fuels has always been "more natural," or at any rate more desirable, for both Canadian fuel consumers and Canadian fuel producers than interprovincial trade in fuels. Second, the marketing of Canadian fuel sources has been a major concern of the federal government by virtue of the fact that such marketing on any large scale has had to involve either international or interprovincial trade and transportation. Third, fuels have figured more prominently and controversially in Canadian energy policy than electrical power, where hydroelectric and thermal-electric production facilities were developed or constructed under the jurisdiction of the two major energy-consuming provinces.) Finally, Canadian fuel policies have been heavily influenced by the fuel import and export policies of the United States.

In fact, given that the geographic circumstances affecting Canadian energy policy have been constant, the case can be made that the single most important variable affecting Canadian fuel policy has been the fuel trade policies of the United States. Many of the measures taken by federal governments to promote Canadian consumption of Canadian fuels have come about as a response to American policies that have either restricted the access of Canadian fuel producers to American markets or threatened the reliability of American fuel supplies to Canadian consumers. In this sense, Canadian proximity to the United States and Canadian dependency upon the United States for both markets and supplies have been highly significant environmental determinants of Canadian fuel policies.

The Distribution of Power

It must be acknowledged at the outset that there is little direct evidence presented in this study as to what groups and individuals have had the greatest access to and influence over the various ministers and prime ministers who have made the decisions reported here. Such knowledge could only come, if at all, from highly detailed studies of the decision-making process leading to the adoption of individual policies. However, to the extent that power can be inferred from results, and to the extent that "the pattern of policy [reflects] the distribution of power and influence," it is possible to draw a few conclusions from the evidence available here about "where the power lies" in Canada with respect to energy policy.[5]

Given the concentration of population and industrial and financial might in the central Canadian provinces, it is striking to observe the frequent success the less populous fuel-producing provinces have had in achieving a measure of protection in the Canadian fuel market for their generally more costly supplies. This success would seem to indicate either that fuel producers have exercised disproportionate power over the federal government, or that voters and interest groups in the consuming provinces have recognized the importance of supporting the fuel industries of the country. At any rate, the fuel-producing provinces have all benefited at one time or another at the expense of at least part of central Canada, although the federal government has consistently balked at imposing the larger burden on the nation's taxpayers (the majority of whom are located in the central provinces) that would have been necessary to bring about a nationwide market for Canadian fuels. Nevertheless, the concentration of political power in central Canada, particularly Ontario, seems to be reflected in the fact that federal measures to expand the Canadian market for Canadian fuels have gone farthest when imported fuels have been unavailable, less reliable, or more costly than Canadian fuels. Canadian producers have been left, on occasion, with considerably less than adequate markets; central Canadian consumers have not been without help from the federal government in gaining access to Canadian supplies when these have been needed. It is true that the federal government took a long time to respond to pressures from both Alberta and central Canada to bring Alberta coal to Ontario during a supply crisis in the 1920s, but there the overriding considerations appeared to be the inordinate cost involved, the realization that crises do not last forever, and an awareness that the solution to the coal difficulties of Ontario lay in a switch from American anthracite to American bituminous coal rather than in a switch from American coal to Alberta coal.

The politics of energy in Canada during the 1970s more clearly revealed the dominance of the central provinces, in that the pricing policies adopted by the federal government benefited the consuming provinces at the expense of the producing provinces; the producers and their provincial governments forfeited billions of dollars in income that would have accrued to them if Canadian price levels had been set closer to the world price. However, many powerful interests located in the central provinces had begun to concede the importance of moving to the world price at the end of the decade, including the Canadian Chamber of Commerce and the Ontario Manufacturers' Association, and many observers began to expect a softening of opposition to higher prices from the government of Ontario following its re-election in 1981. Indeed, Canadian prices were scheduled to rise more rapidly during the 1980s than the federal Liberals had previously promised in the national election of 1980, which could be taken as evidence of a shift in the balance of political forces in favour of the producing provinces or, alternatively, as a change in the view of their own self-interest held by powers-that-be in central Canada.

This problem of interpretation is one that plagues all attempts to infer power from results—that is, to determine who holds the greatest power from evidence showing whose interests were best served by what governments have done. The "self-interest" of the powerful is not necessarily or uniformly identical with their immediate, short-term interest or, strictly speaking, even their economic interest. This study has shown that, to a considerable extent, the policies adopted by the federal government concerning the size of the Canadian market for Canadian fuels do appear to have represented a balance struck between the economic interests of the consuming and producing provinces; but they have also represented a balance struck between the central Canadian interest in the most economic source of supply of its fuels and its interest in the much broader, less directly economical goals of national independence and national unity. It is one thing to say that governments do what the powerful want, or at least do not oppose; it is another to assume that the powerful are motivated by nothing more with respect to Canadian energy policy than obtaining energy supplies at the lowest possible prices.

The role of such countervailing ideas and goals in shaping Canadian fuel policies will be explored in the next section. Before leaving this rather speculative discussion of the distribution of power as it affects Canadian energy policy, however, one conclusion can be stated with some confidence. The analysis presented in this study lends no support to the tirelessly repeated view that Canadian energy policy—especially as it relates to the question of international as opposed to national markets for Canadian oil and natural gas—has been and is almost exclusively the result of the high level of foreign control in the oil and gas industry. It should be acknowledged that this view is widely held for a very good reason: foreign-controlled companies, as this study and several others have amply demonstrated, have been heavily involved in the formulation and implementation of Canadian policies in this area, probably more heavily involved than all other sources of influence and information combined. This study casts doubt not upon the proposition that Canadian policies have conformed closely to the expressly preferred policies of foreign-controlled petroleum companies, but rather on the *assumption* that these policies are promoted by the companies involved *because they are foreign-controlled.* This study has demonstrated that Canadian fuel producers—regardless of the fuel, regardless of the region involved, regardles of the period, and regardless of the location of ownership of the industry—have preferred to sell to markets located in the United States rather than to the central Canadian market, mainly because those U.S. markets have been more lucrative owing to the lower transportation costs associated with them. This is not to say that foreign control and vertical integration of the oil and gas industry has not been a factor in other matters such as domestic oil pricing (prior to 1973) or gas export pricing; but it is to point to an industrial or sectoral interest in many areas of policy that is common (or could be expected to be common) to both Canadian- and American-owned firms.

It can be inferred from this that a Canadian petroleum industry owned entirely by Canadians would have called for policies not significantly different from those supported historically by the foreign-controlled producers. Logic aside, it is worth remembering that many of the Canadian and independent firms who advocated an Alberta-Montreal oil pipeline in the late 1950s, when the foreign-controlled firms opposed it, did so only if it proved impossible to negotiate greater access to the American market. This stance had a historical precedent in the fact that support from Nova Scotia for a tariff against coal imports from the United States was often based on a desire to negotiate the removal of the American tariff against coal imports from Canada, in the service not of protectionism but of unhindered fuel trade between the two countries. From a policy standpoint, the relevant fact about foreign-controlled fuel producers is that they are producers, not that they are foreign-controlled. This is not to say—or to deny—that the policies they have advocated as producers have been in the national interest; it is to say that their preferred policies—and, hence, most of the time, the policies of the government of Canada—have not necessarily or incontestably been the result of foreign control in the industry. The historical record reported here suggests that if a foreign influence is to be observed in Canadian energy policy, it is in the trade policies of American governments, not in the desire of foreign-controlled firms to sell their product in its most remunerative market, a trait that might also be expected in Canadian-owned firms, public or private.

There are many other examples to support this contention that the location of ownership of firms has nothing to do with their preferred policies with respect to the marketing of Canadian fuels. Both foreign- and Canadian-controlled firms showed the same interest in disposing of Alberta surpluses in the late 1970s by means of increased natural gas exports to the United States, and were equally opposed to the idea that surplus volumes of gas ought to be retained in Canada for future Canadian use in eastern markets. Finally, the federal government's own oil company, Petro-Canada, is involved in several energy projects that include provisions for the export of Canadian resources.[6] Apart from demonstrating that a reluctance to be confined to Canadian markets has nothing to do with the ownership—Canadian or American, public or private—of firms, this last example seems to suggest that in the absence of a clear mandate from government concerning priorities to be pursued in the national interest, public ownership in an industry has no advantage over the regulation of an industry as a method of directing the efforts of that industry toward stated national objectives. Either that, or it suggests that the desire to promote energy exports is simply a matter of the economics of the industry.

Ideas

The notion that ideas tend to account for "broad orientations rather than the specific details of policy" seems well borne out in the area of fuel policy.[7] Debates relating to the marketing of Canadian fuel production and the acquisition of fuel for Canadian markets have consistently involved appeals to some of the fundamental ideas and principles informing the country's political life, although it would be misleading to suggest that the acquisition of fuels has consistently been a central issue in Canadian politics. Despite its persistence as a problem, and despite the several occasions when the problem has grown into a crisis, it can hardly be said that the quest for reliable, low-cost supplies of fuel has been a driving force in Canadian history. Nevertheless, it is fair to say that the approach taken to Canada's fuel problem during different periods in its history has reflected basic attitudes current to the time, so that the history of Canadian fuel policy reads like a minor theme in the story of the country itself. Coal policy was debated as a matter of national unity and nation-building before World War I and as a matter of national independence between the wars; oil and gas policy, particularly with respect to the construction of pipelines during the years following World War II, was debated in terms that reflected the country's ambivalence between the desire to keep "Canada First" and the desire to reap the potential benefits of closer North American cooperation. Canada's decision to proceed with the Alaska Highway pipeline, which has been aptly described as an act of generosity to our American friends that has nothing to do with Canadian energy policy as such, can be taken as an indication of the present state of the balance between these two aspirations, as well as the culmination of two decades of decisions regarding oil, natural gas, and pipelines that were expressly directed toward the creation of a continental fuel market or, at the very least, dedicated to the preservation of "amity and comity" in relations between Canada and the United States.

This correspondence between the fundamental orientation of Canadian fuel policy and the more general orientation of national politics is neither coincidental nor purely conceptual since the central issue in Canadian fuel policy—the extent to which the federal government ought to intervene in order to expand Canadian consumption of Canadian fuels—has never been addressed exclusively in economic terms, and has therefore always involved considerations of some overriding national purpose or national interest that could justify the inevitable departure from the most economically rational solution. The linking of Canadian sources of supply and major Canadian markets for fuels has always been (as one observer put it with reference to Alberta coal during the 1930s) an element of the "willed economy" rather than the "natural economy" of the country, and for that very reason Canadian fuel policies at different times are indicative of, because they were derived from, the prevailing sense of national purpose and the determination of the country to remain united and independent. Variations in Canadian fuel policy have therefore been the product of variations in the extent and intensity of the determination, in particular, to establish and maintain the facilities and policies necessary to provide Canadian fuel producers with access to Canadian markets, and Canadian consumers access to Canadian supplies. This determination has been, in turn, a product of conflicting interests among Canadians in energy matters and of the circumstances prevailing in continental and world fuel markets.

The point has been made earlier that the most obvious explanation for Canada's failure ever to have realized a national fuel policy whereby all Canadian markets were served with Canadian supplies is simply its excessive cost. The political advantages of a fully "national" fuel market have never appeared large enough to warrant their economic price. But this raises the question of why successive Canadian governments have seen fit to restrict imports into Canada at all and to otherwise assist Canadian fuel producers in expanding the Canadian market for their production. The obvious answer, and one—perhaps the only one—which continues to be stressed in current discussions of energy policy, would seem to be the desire of the country as a whole to be self-sufficient in fuels so that it would not be vulnerable to embargoes by foreign suppliers who might attempt to change Canadian foreign policy, to alter the balance in negotiations in other areas, or simply to enforce higher prices. However obvious this answer may appear, particularly in the context of the events since the OPEC price hike of 1973, it is probably inaccurate and is certainly inadequate.

The logic of the "self-sufficiency as insurance" position would seem to entail not partial but total self-sufficiency—the vulnerability to blackmail would not disappear if one-half or one-third of the country was exposed to the threat of disaster—and, as we have seen in the past and continue to see in the present, the Canadian government has never taken action to create a capacity for absolute self-sufficiency. It has instead acted to preserve some limited *portion* of the Canadian market for Canadian producers. The social and economic dislocations conjured up by those advocating self-sufficiency as a safeguard against deliberate suspensions in deliveries of foreign fuels would occur very quickly, or at any rate soon enough that any part of the country left exposed to them in normal times probably could not be saved from them in a time of crisis. Such uncertainty is a strange component to discover in a policy whose rationale is obstensibly the avoidance of risk. For another thing, to put logical considerations to one side in favour of the historical record, the only two occasions on which the idea of a national fuel policy received widespread political support and some immediate, though only partial, government action have followed *actual* (and fairly severe) dislocations in the availability and price of imported fuels: coal imports in the 1920s and crude oil imports in the 1970s. This seems to suggest that the primary justification for the national fuel policy from the consumer standpoint and from the standpoint of the national interest has not been the absolute one of the protection of Canadians in every part of the country against the sudden *unavailability* of foreign supplies, but rather the relative one of reducing the impact on the national economy of significantly higher *prices* for these supplies. These considerations lend support to the view that the desire for self-sufficiency *as such* has not produced government action aimed at enlarging the Canadian market available to Canadian fuel, but that the desire to gain greater access to Canadian supplies when they were cheaper than imported supplies has prompted moves in that direction. It is true that C. D. Howe actively promoted the Trans-Canada pipeline as a means of providing Ontario with secure access to an increasingly important fuel source in light of the fact that imports from the United States could not be relied upon, but even that project was not part of any general strategy of self-sufficiency in fuels, as continued imports of coal into Ontario and of overseas oil into Quebec attested at the time.

Having put aside full self-sufficiency as an objective of or rationale for steps taken toward a national fuel policy, we still need to know why different Canadian governments have seen fit on occasion to expand Canada's consumption of its own fuels despite the availability of lower-priced imports. Here, the answer would appear to be the desire (or at least the willingness) of nearly all Canadian governments since Confederation to keep Canada's fuel industries in production in order to prevent the economic collapse or stagnation of the provinces in which they have been located. Since access to U.S. markets has played a part in the expansion of all Canadian fuel industries, subsequent American restrictions on Canadian access to those same markets have been a frequent cause of severe difficulties (shut-in production) for those same industries and have, therefore, triggered pressures from producing provinces and the fuel industries for assistance from the Canadian government in capturing a larger share of the Canadian market. Just as Confederation itself was largely conceived by its supporters as compensation for the inability of the Canadian provinces to achieve reciprocity with the United States, steps toward a national fuel policy have frequently been taken in response to a failure to achieve or renew access to American fuel markets, either through the elimination of American tariffs or quotas on imports from Canada or through some sort of "swapping" arrangement. It may or may not be true that the renewal of reciprocity with the United States at some point after 1867 would have meant the abandonment of Confederation; but it seems almost certain that completely unrestricted access to the American fuel market would have meant the abandonment of national fuel policies. Moreover, three of the attempts by federal governments in the direction of a national fuel policy—the attempt to increase the Canadian consumption of Canadian coal in Ontario in the 1920s, the construction of the Trans-Canada pipeline, and the extension of the Interprovincial pipeline from Sarnia to Montreal—occurred under conditions where Canadian officials had asked the American government to provide them with assurances of long-term supplies of the particular fuel source involved and were unable to obtain them. It may be that the governments of Canada have pursued a national fuel policy to the precise extent of their failure to achieve a continental one.

At any rate, there is ample evidence that Canadian governments have been keenly interested in Canadian-American cooperation in energy matters whenever interest in them has been evident on the American side and whenever they have not seemed inimical to the interests of the central Canadian provinces. Canada still relies on imports of American coal rather than Canadian coal; some "American connection" is part of every major Canadian transmission system ever built in the country; the "pre-build" portions of the Alaska Highway pipeline and its attendant gas exports were approved before the Trans-Quebec Maritime pipeline and the expansion of gas service east of Montreal were approved; the Alaska Highway project itself remained a national priority despite the lack of any obvious contribution to Canada's energy position; and the marketing of the oil and gas in the frontier areas that were so favoured in the national energy program is almost bound to require another round of Canadian-U.S. cooperation in the transportation and marketing of Canadian fuel. However, as far as such policies have stemmed from any fundamental notions and orientations, they would seem to have had less to do with an explicit commitment to continentalism as such than with a lack of commitment to exclusively Canadian solutions to Canadian energy problems. Evidently nothing about Canada is generally perceived to be placed in jeopardy by North American interdependence in energy matters: neither Canadian unity nor Canadian independence are deemed to require the complete substitution of interprovincial trade in fuels for international trade in fuels (except, of course, when international trade arrangements are "not on" owing to American reluctance to supply Canadian needs on a firm basis or to open the American market to Canadian producers). Canada prospers when the continent works as one country. Self-evident or not, this proposition seems to have governed Canadian thinking about fuels.

Institutions and Processes

This study has not included the kind of cross-national comparisons of energy policies that would be most conducive to conclusions about the role of institutional structures such as federalism in the formulation of Canadian policies toward the marketing of Canadian fuels, or the kind of detailed examination of individual decisions that would support generalizations about the importance of the processes by which such decisions are arrived at.[8] Nevertheless, the material presented here does permit a few observations concerning the extent to which the other three determinants examined here—environment, power, and ideas—appear likely to limit or mitigate the *independent* effect of these remaining two types of determinants. For example, the powerful and consistent influence of regionalism (or of centre-periphery relations) in the politics of fuels in Canada seems very likely to limit—indeed, to subsume—the distinct contribution to outcomes of, say, the division of powers between levels of government or political partisanship. Since anything other than a tentative and selective discussion along such lines would require at least one volume the size of the present one, these concluding remarks will merely indicate the direction a comprehensive analysis might take.

As regards the seemingly paramount institutional reality of Canadian politics, the division of powers between province and central government, it might seem that in the absence of changes over time in the relevant constitutional arrangements, little can be said about it as a variable affecting Canadian energy policies. However, it is worth noting that some variation has occurred over time with respect, at least, to the constitutional provisions concerning Alberta's fuel resources. The coal crisis of the early 1920s occurred before the control of natural resources was transferred from the federal government to the province. Compared with more recent developments in the 1960s and 1970s, this does not seem to have resulted in any crucial differences on at least two scores: producers in Alberta still saw the Alberta government as the first line of defence (or assault) on the federal government, whose policies they perceived to be insensitive to the plight of the industry; and the fact that control of the resource resided with the federal government did not appear to add any incentive on its part to promote the interests of the industry. If this is enough to suggest that perhaps geography and regional interests are more fundamental to national energy policies than is the constitution, it can also provide some basis for speculation about what energy policy and energy politics might have looked like if Canada had been a unitary state, controlled from the Golden Horseshoe of central Canada.

There seem to be grounds to assume that the regional antagonisms and conflicts of interest over the question of the proper disposition of Canada's fuel supplies observed in the preceding chapters would still have arisen in a unitary Canadian state, except in so far as recent federal-provincial and interprovincial disputes over fuel trade and marketing policies have been confused with the more intense conflict between governments over revenue sharing and the distribution of resource rents. Under the same environmental constraints of geography and American trade policies, the regional distribution of costs and benefits associated with the possible creation of a national fuel market would not be substantially different, and would produce similar political consequences. Trade and transportation policies can benefit one region at the expense of another in a unitary state as well as in a federal one, and interregional conflicts over such policies need not be any less volatile than interprovincial or federal-provincial conflicts. The British experience with North Sea oil development and the concomitant rise of Scottish nationalism attests to the fact that unitary constitutional arrangements do not, in and of themselves, prevent regional conflicts of interest and centrifugal political tendencies from intensifying over issues relating to the disposition of energy resources, especially when production and consumption are concentrated in separate regions with highly unequal populations.[9]

As regards the independent contribution of process variables to the trade and transportation policies of the federal government, this study has little to offer, although the material presented in the foregoing chapters does support a few observations. One has to do with the role of Royal Commissions and regulatory agencies, as opposed to Parliament, in the deliberation and formulation of Canadian fuel policies, and another has to do with the lack of clear and significant differences between the approaches taken to these matters by the Liberal and Progressive Conservative parties. Again, the argument here is that these factors are explained by or subsumed under the variables already examined in this study. The fact that neither political party can govern for long if its policies do not conform to the economic interests of the voters in the central provinces, when combined with the regional configuration of interests on the question of fuel supply, probably accounts for the fact that in actual approach (though not, at times, in rhetoric) the two major parties have not acted very differently when they have had the power and the responsibility to act in these matters. Similarly, it may be that in so far as any action with respect to the marketing of fuels in Canada does entail such an asymmetrical regional distribution of costs and benefits, the starker political conflicts over the distributive consequences of energy policy is best played out in the more remote and obscure arenas of commissions and regulatory agencies, where the setting of freight rates and the estimation of gas surpluses can translate issues of economic justice (or injustice) into a less divisive debate among experts seeking to discover the "unique solution" to a technical problem.

In sum, this discussion rests on its earlier conclusion that the fundamental condition affecting Canadian fuel policies is that of geography translated into transportation costs, and that the most significant variable has been the fuel import and export policies of the United States. As Thomson recognized, the persuasiveness of this conclusion can be enhanced if one contemplates the patterns of fuel transportation and trade that one might have observed if Canada and the United States had been one country.[10] Alternatively, and possibly more persuasively, one might ask how this volume would have read if, despite their status as distinct countries, the first Canadian coal tariff had indeed brought about the elimination of the American one and had ushered in a century of unrestricted fuel trade between the two continental partners.

Notes

1. The Quest for a National Fuel Policy

(For the reader's convenience, shortened references are used in the notes to the chapters. Complete publication information for all works cited will be found in the bibliography.)

1. Evans, *Western Energy Policy*, ch. 2.
2. Lindberg, "Energy Policy and the Politics of Economic Development," pp. 357-8.
3. Dales, "Fuel, Power and Industrial Development in Central Canada," pp. 185, 196.
4. Nelles, *The Politics of Development*, chs. 6 and 7, esp. pp. 216-19, 291.
5. Canada, House of Commons, *Debates,* June 13, 1946, p. 2449. Hereafter cited as *Debates*.
6. H. A. Innis and Lower, *Select Documents, 1783–1883*, p. 6.
7. See, for example, Nordeg, *The Fuel Problems of Canada*, ch. 7. Of course, a similar situation prevails at present with respect to oil and gas, the bulk of which lies in Alberta.
8. L. R. Thomson, "Some Economic Aspects of the Canadian Coal Problem," pp. 403-4.
9. For an analysis of the superior economics of continental as opposed to national trade in natural gas, see Waverman, *Natural Gas and National Policy*.
10. Charles Tupper, *Debates*, Feb. 24, 1876, p. 231. The "tax" in question was a tariff on imports of American coal.
11. See, for example, H. A. Innis, "Transportation as a Factor in Canadian Economic History," pp. 186-99.
12. Taken from Davis, *Canadian Energy Prospects*, table 1 of app. B, p. 348.

2. Coal and National Unity, 1867–1913

1. Davis, *Canadian Energy Prospects*, p. 319; Royal Commission on Coal, 1946, *Report*, p. 379 and app. C. Coal increased its share of total energy consumption largely through the displacement of wood.
2. Brown, *The Coal Fields and Coal Trade*, ch. 5.
3. Ibid., pp. 47-48.
4. There is, nevertheless, an indication of some commercial sales of Cape Breton coal as early as 1767. On June 18 of that year a notice appeared in the *Quebec Gazette* advising of Spanish River coal for

sale to ships landing there. See H. A. Innis, *Select Documents, 1497–1783*, p. 213.

5. Brown, p. 54.
6. H. A. Innis, *Select Documents, 1497–1783*, pp. 14, 212.
7. Brown, pp. 66-7.
8. Ibid., p. 91.
9. Forsey, *National Problems of Canada*, p. 5. Cf. Davis, pp. 79-80, where, at least implicitly, "free entry" is regarded as responsible for the high level of exports to the U.S. in the mid-1860s.
10. Davis, p. 80.
11. See Creighton, *The Road to Confederation*, and Waite, *The Life and Times of Confederation*.
12. Province of Canada, Legislature, *Parliamentary Debates on the Subject of the Confederation of the British North American Provinces*, Feb. 7, 1865, pp. 64-5.
13. Ibid., Feb. 9, 1865, pp. 141-2.
14. Ibid., Feb. 20, 1865, p. 355.
15. Ibid., March 10, 1865, p. 953.
16. Ibid., Feb. 17, 1865, p. 280.
17. Ibid., March 13, 1865, p. 1029.
18. Canada, House of Commons, *Debates*, April 7, 1870, pp. 931-40.
19. Ibid., April 26, 1870, pp. 1192-9.
20. Ibid., April 27, 1870, pp. 1223-6.
21. Ibid., May 3, 1870, pp. 1329.
22. Ibid., March 22, 1871, p. 598.
23. Ibid., April 26, 1870, pp. 1191-1214; April 27, 1870, pp. 1221-63; March 22, 1871, pp. 586-98.
24. To place the issue of protection for the Canadian coal industry in the more general context of the politics of Canadian tariffs, see Acheson, "The National Policy and the Industrialization of the Maritimes," and Foster, "The Coming of the National Policy."
25. *National Problems of Canada*, p. 6.
26. See Muise, *Elections and Constituencies*, pp. 13, 31, 79, 407. In this study, Cumberland is included among the regions of Nova Scotia whose economic orientation toward "coal, iron and rails" led them to greater enthusiasm for Confederation in the politics of the province, as compared with those regions whose opposition to Confederation was partly based on their economic orientation toward the traditional "wood, wind and sail." This distinction and its political significance are carried forward after 1867 in the "protectionist" versus "free trade" debates. H. A. Innis also describes Tupper as having "effectively represented the coal interests" of Nova Scotia through his support for the National Policy and

associated developments in "The Canadian Mining Industry" in Mary Q. Innis, ed., *Essays in Canadian Economic History*, p. 313.

27. Forsey, app. A; Royal Commission on Coal, *Report* (1946) table p. 64.
28. *Debates*, Feb. 24, 1876, pp. 225-6.
29. Ibid., p. 232.
30. Ibid., p. 231.
31. Ibid., 229. A report of a committee of the House of Commons at the time cited evidence to the effect that the price of American coal in Toronto was in the range of $3.40 to $4.80 per short ton, an indication that the $2.00 difference between the price to Ontarians of Nova Scotian coal and American coal would not be inconsequential. See document reproduced in Burley, *The Development of Canada's Staples*, p. 185.
32. *Debates*, Feb. 13, 1877, p. 39.
33. Ibid., March 1, 1877, p. 379.
34. Ibid., pp. 381-3.
35. Ibid., p. 391.
36. Ibid., p. 381.
37. Ibid., March 7, 1877, pp. 545-6.
38. House of Commons, *Journals*, vol. XI, app. 4, 1877, pp. 68-78.
39. Ibid., p. 15.
40. Ibid., p. 18.
41. Ibid., p. 61.
42. *Debates*, March 14, 1879, p. 476.
43. Ibid., April 22, 1879, pp. 1426-37.
44. Ibid., p. 1430. Cf. Paterson, p. 1434.
45. Forsey, p. 120.
46. Acheson, "The National Policy and the Industrialization of the Maritimes," p. 20.
47. Forsey, p. 7.
48. Royal Commission on Coal, *Report* (1946), p. 65.
49. *Debates*, January 28, 1884, p. 65.
50. Ibid., June 1, 1884, p. 3571.
51. Ibid., March 22, 1909, pp. 3092-4.
52. Royal Commission on Coal, *Report* (1946), p. 64.
53. *Debates*, March 22, 1909, pp. 3106-3108.
54. See Fowke, "The National Policy—Old and New," pp. 271-86, where he distinguishes between the earlier tasks of the federal government involved in the promotion of national development and the later ones involved in the provision of public welfare.
55. Royal Commission on Coal, *Report* (1946), pp. 433-4.
56. See Forsey, pp. 120-1.

3. Coal and National Independence, 1919–1946

1. Royal Commission on Coal, *Report* (1946), p. 66, hereafter cited as R.C.C. *Report*.
2. This report contains a good summary of the regulation of the Canadian coal market during the war. See Canada, *Final Report of the Fuel Controller,* pp. 15-35. See also R.C.C. *Report*, pp. 528-9.
3. *Final Report* of the Fuel Controller, p. 41. Evidence within the report would seem to support this conclusion.
4. R.C.C. *Report,* p. 529.
5. *Final Report of the Fuel Controller*, p. 69.
6. Ibid., italics added.
7. Ibid., 37.
8. Ibid.
9. Letter from Hon. H. Greenfield, Premier of Alberta, to the Rt. Hon. Mackenzie King, Prime Minister of Canada, May 2, 1923. Provincial Archives of Alberta, Premiers' Papers, file 322. Hereafter cited as Premiers' Papers.
10. R.C.C. *Report*, p. 68.
11. Quoted in C. J. Broderick, *Coal*, 1931. Premiers' Papers, file 807.
12. R.C.C. *Report*, pp. 444-53.
13. Ibid., pp. 71-2.
14. Ibid., p. 71.
15. Ibid., pp. 446-8.
16. An interesting picture of the problems and prospects of the Alberta coal industry immediately after World War I is available in Bercuson, ed., *Alberta's Coal Industry, 1919*, which provides an edited transcript and the report of the Alberta Coal Mining Commission of 1919.
17. R.C.C. *Report,* p. 530.
18. See Burley, *The Development of Canada's Staples*, p. 198.
19. R.C.C. *Report*, p. 568.
20. For a comprehensive breakdown of federal measures in aid of Canadian coal movements, see ibid., ch. 7.
21. Ibid., p. 439. Its share of the market jumped by 5 percent between 1931 and 1933.
22. Ibid., p. 581.
23. Obituary in the Toronto *Globe*, Feb. 8, 1950; the Toronto *Telegraph*, Dec. 19, 1921.
24. *Debates*, March 31, 1924, p. 842.
25. Ibid., March 10, 1920, p. 322.
26. Ibid., pp. 322-3.
27. Ibid. It may be worth recording for the sake of those who believe there is occasionally something new under the sun that, at this

point, Burnham turned his attention from coal to "industrial alcohol" as an alternative fuel available to Ontario, which he asked the premier of Ontario to consider.

28. See ibid., pp. 329-30. Asked if it was not a fact that "the cost of freight and handling alone" in bringing coal from Edmonton to Toronto "would be greater than what we have to pay now for bituminous coal from [the United States]," Meighen replied, "I think it is."
29. Ibid., March 23, 1921, p. 1247.
30. Ibid., p. 1269.
31. Ibid., March 19, 1923, pp. 1262-5.
32. For a discussion of some of these difficulties, see R.C.C. *Report*, pp. 362-7.
33. Ibid., pp. 350-51.
34. Reproduced in Burley, p. 193.
35. *Debates*, March 19, 1923, p. 1269. Elsewhere on the same page another document is cited to the effect that the current rate was $12.50 per ton.
36. Ibid., p. 1280.
37. See Canada, Special Committee of the House of Commons, *Evidence Taken Respecting the Future Fuel Supply of Canada*, 1921, pp. 102: 167-71; 394-8. For an excerpt from the report of this committee, including its recommendations, see Burley, pp. 186-9.
38. Canada, *Third and Final Report of the Select Committee on Mines and Minerals* (Ottawa, 1923, pp. 7-9). Reproduced in Burley, pp. 198-9.
39. Canada, *Final Report of a Special Committee on the Fuel Supply of Canada* (Ottawa, 1923), as reproduced in Burley, p. 191.
40. *Debates*, March 31, 1924, pp. 842-9.
41. Ibid., p. 886.
42. Ibid., p. 885.
43. Ibid., p. 876.
44. Ibid., pp. 856-7.
45. Hopkins, *The Canadian Annual Review of Public Affairs*, 1922, pp. 118-20. A long list of White's publications appears as a footnote to p. 119.
46. Ibid., 1923, p. 747.
47. See Premier's Papers, files 322 and 323.
48. Ibid., Aug. 21, 1923, file 322.
49. Ibid., Sept. 28, 1926, file 323.
50. Howard Stuchbury to Premier Greenfield, Premiers' Papers, Dec. 2, 1923, file 322.
51. Ibid., Dec. 16, 1923, file 322.
52. Ibid., July 10, 1925, file 323.

53. Resolution passed March 19, 1923, enclosed in a letter from Green-field to King, ibid., May 2, 1923, file 322.

54. P.M.O. to Greenfield, Ottawa, ibid., May 23, 1923.

55. Letter to Greenfield, Ottawa, ibid., May 19, 1923.

56. Ibid. The W.L.M. King Papers in the Public Archives of Canada generally confirm the impression of federal indifference on this question between 1923 and 1925. See particularly vols. 76 and 101.

57. Premiers' Papers, May 4, 1925, file 323.

58. Ibid., July 14, 1925, file 323.

59. Telegram, Greenfield to Ferguson, ibid., Feb. 27, 1924, file 322.

60. See *Debates*, March 16, 1925, pp. 1195-1203; March 15, 1926, pp. 1572-93; Feb. 16, 1927, pp. 371-83; July 15 and 16, 1931, pp. 3797-3820 and 3831-53; March 7, 1934, pp. 1261-72; May 22, 1939, pp. 4357-74; and June 13, 1946, pp. 2443-57.

61. See William W. Goforth, *The Case for a Coordinated Fuel Policy*, undated pamphlet (probably 1928), p. 3: Premiers' Papers, file 324, item 1100-12-3.

62. *Debates*, Feb. 16, 1927, p. 381.

63. Thomson, "Some Economic Aspects of the Canadian Coal Problem," p. 403. The Royal Commission on Coal, 1946, placed the same figure in the order of $100 million: *Report*, p. 582.

64. *Canada Year Book*, 1927-28, p. 811.

65. Goforth, p. 8.

66. The Royal Commission on Coal was struck in October 1944, following a report to the Privy Council from the Minister of Munitions and Supply. The problems with Canada's fuel supply during World War II are reviewed at length in the Commission's report at pp. 532-63. Generally, both imports and Canadian production increased during World War II, as did government controls relating to financing, pricing, allocation among users, and labour relations. The commission's terms of reference were comprehensive, as was its report.

67. See the submission to the Royal Commission on Coal by the Associated Boards of Trade of Cape Breton Island, January 16, 1945: Public Archives of Canada, *Royal Commission on the Coal Industry in Canada, Exhibits*. See also the brief of the Western Canada Bituminous Coal Operators Association, April 1945, Exhibits, vol. 23, no. 117, pp. 45-6.

68. See, for example, the submission on behalf of Canadian Collieries (Dunsmuir) Ltd., R.C.C. *Report* (Exhibits), vol. 22, no. 92, passim.

69. See, for example, brief submitted by District No. 18, United Mine Workers of America, Calgary, April 1945, ibid., vol. 23, no. 122,

pp. 10-13; memorandum submitted on behalf of Western Canada Fuel Association, ibid., vol. 25, no. 158, p. 16.

70. Statement of the Position of the Government of Ontario Before the Coal Commission, August 1945, ibid., vol. 25, p. 169.

71. See, for example, Submission of the Toronto Coal Exchange, ibid., vol. 28, no. 229, pp. 50-51; brief presented by Fuel Protective Association of Hamilton, ibid., vol. 26, no. 183, p. 2; letter from Mongeau & Robert Cie, Limitée, Aug. 13, 1945, ibid., vol. 26, no. 193.

72. R.C.C. *Report*, p. 582.

73. Ibid.

4. Nationalism Versus Continentalism: Oil and Natural Gas Pipelines, 1949–1958

1. Canada, Department of Defence Production, *The Natural Gas Industry*, p. 37n.

2. *Debates*, April 8, 1949, p. 2509. Parts of this chapter draw heavily on material contained in Askew, *Continentalism Versus Nationalism*, ch. 2.

3. *Debates*, April 8, 1949, p. 2509.

4. Ibid.

5. Ibid.

6. Ibid., April 27, 1949, pp. 2642-3.

7. Ibid., April 8, 1949, p. 2512.

8. Ibid., June 13, 1946, p. 2451.

9. Alberta, Royal Commission on Alberta's Oil Industry, *Report* (1946), p. 71.

10. See statement of the Minister of Reconstruction and Supply (Right Hon. C. D. Howe), *Debates*, May 28, 1947, p. 3514. This statement was to the effect that Ontario refiners felt, on being consulted by the minister, that they would be short of fuel oil, not only to meet the demand of new customers but to continue to serve present customers.

11. *Debates*, Jan. 27, 1948, p. 592; Feb. 13, 1948, pp. 1194-7. For a brief discussion of Canadian-American relations on the matter of American oil exports to Canada, 1947-48, see Page, "An Energy Crisis in Reverse." It is clear from this account that despite congressional pressures on them to suspend exports to Canada, U.S. officials gave precedence to the "spirit of Hyde Park" over a narrow view of America's economic self-interest.

12. *Debates*, April 30, 1949, pp. 3808-9.

13. Hanson, *Dynamic Decade*, p. 240.

14. Aitken, "The Changing Structure of the Canadian Economy," p. 23.
15. *Debates*, Sept. 22, 1949, p. 159.
16. Ibid., Oct. 4, 1949, p. 520. The importance of prospective exports to the United States to the routing of the pipeline through Superior, and later south of the Great Lakes, is difficult to establish with certainty. On the one hand, the Great Lakes states seemed one of the more promising future markets for Alberta's growing production. On the other hand, the Toronto market appeared to offer producers a higher rate of return. The issue is further confused by American import restrictions, as well as by the supply preferences of the major international parents of the Canadian companies who sponsored Interprovincial. See Hanson, pp. 154-6; Davis, *Canadian Energy Prospects*, pp. 124-6; Plotnick, *Petroleum*, pp. 61-8.
17. Debates, Oct. 4, 1949, p. 522.
18. Ibid.
19. Ibid., pp. 525-6.
20. Canada, Board of Transport Commissioners, *Transcript of Evidence*, vol. 821, pp. 5735-89. Hereafter cited as BTC Hearings.
21. Ibid., pp. 5753, 5756. The Interprovincial hearings included an interesting argument concerning the connection between exports and the American route for the pipeline. The chairman of the commission expressed concern at one point that a possible difficulty with the U.S. route might be that the U.S. government could impose undesirable conditions on the construction of the line, and went on to speculate that perhaps such a development would be less likely if the company were to obtain outlets in the United States itself, an assessment with which the company representative concurred.
22. *Debates*, Oct. 28, 1949, pp. 1247-8.
23. Ibid., Nov. 11, 1949, p. 1968.
24. Ibid.
25. See, for example, ibid., Nov. 1, 1949, pp. 1348-9; Nov. 8, 1949, pp. 1808-9; Dec. 6, 1949, pp. 2808-9.
26. Ibid., Nov. 8, 1949, p. 1558; Nov. 26, 1949, p. 2430.
27. See ibid., Nov. 2, 1949, pp. 2101-2, for the suggestion of a single line in each direction.
28. See ibid., Nov. 1, 1949, p. 1349; Nov. 8, 1949, p. 1806.
29. Ibid., Nov. 29, 1949, p. 2428.
30. Ibid., Nov. 8, 1949, p. 1558.
31. Ibid., April 29, 1949, p. 2779.
32. Thorburn, "Parliament and Policy-Making," p. 518.
33. BTC Hearings, Sept. 21, 1949, vol. 821, p. 5802.
34. Ibid., pp. 5830-1.
35. The Dinning Commission. See Alberta, Oil and Gas Conservation

Board, *Gas Export, 1950-1960*, pp. 2-3. Hereafter cited as *Gas Export*.

36. Ibid., p. 4.
37. The role of this provincial agency in the matters discussed in this chapter has been reviewed in *Gas Export*. Another significant measure adopted by the Alberta government during the early 1950s, in addition to those just mentioned, was the creation in 1954 of the Alberta Gas Trunk Line Company to gather and transmit natural gas within the province to provincial companies and to exporting companies such as Trans-Canada. See *Gas Export*, p. 177ff. For a review of the role that this firm and other Alberta policy instruments played in Alberta's economy and in relation to national policy during this period and later, see Bregha, *Bob Blair's Pipeline*, and Richards and Pratt, *Prairie Capitalism*, chs. 3 and 4.
38. See, for example, *Debates*, June 1, 1951, p. 3644. This was a debate on Champion Pipe Line Corporation, the sponsor of which was congratulated by Howard Green for including this provision in its bill of incorporation.
39. Ibid.
40. Ibid., June 12, 1951, p. 3998.
41. Ibid., March 16, 1951, p. 1339.
42. See letter from C. D. Howe to N. E. Tanner, Minister of Lands and Mines for the Province of Alberta, Sept. 16, 1950: Public Archives of Canada, C. D. Howe Papers, vol. 30, file 23. The letter informed Tanner that Howe had been advised by an official of the United States Munitions Board that "diversion of normal oil supplies to the Far East" had "seriously accentuated the scarcity" of fuels in the American Pacific Northwest.
43. *Debates*, March 6, 1951, p. 969.
44. BTC Hearings, April 4, 1951, vol. 878, p. 2775.
45. Ibid., vol. 892, p. 8939.
46. *Gas Export*, p. 82.
47. Ibid. Cf. Ian McDougall, "The Canadian National Energy Board," pp. 353-4. See also exchange of correspondence between the United States Director of Defense Mobilization, the Canadian Minister of Trade and Commerce, and the Alberta premier, between Feb. 16 and March 20, 1951: Public Archives of Canada, C. D. Howe Papers, vol. 30, file 23.
48. *Debates*, March 13, 1951, p. 1208.
49. Ibid., April 6, 1951, p. 1695.
50. Ibid., April 24, 1951, p. 2344.
51. *Gas Export*, p. 83.
52. *Debates*, March 9, 1951, pp. 1104-5.
53. Ibid., Feb. 27, 1951, p. 736.

54. For an authoritative account of the Trans-Canada debate, see Kilbourn, *Pipeline*.
55. *Gas Export*, p. 96. See also Ian McDougall, "The Canadian National Energy Board," fn. 34. The letter cited in this footnote by McDougall is available in the Public Archives of Canada, C. D. Howe Papers, vol. 29, file 18.
56. See Dexter, "Politics, Pipeline and Parliament"; Kilbourn; and Thorburn, "Parliament and Policy-making," p. 527 and passim.
57. Gas Export, pp. 23-4.
58. Ibid., p. 14.
59. Ibid., pp. 61-2.
60. See Kilbourn, ch. 4. esp. p. 42.
61. See United States, Federal Power Commission, 12 *FPC* (1953), pp. 313-14.
62. *Debates*, March 13, 1953, pp. 2928-9.

5. Nationalism Versus Continentalism: The Borden Commission and the Marketing of Canadian Oil and Natural Gas, 1958–1960

1. Aitken, et al., *The American Economic Impact*, p. 22.
2. *Debates*, Oct. 15, 1957, pp. 12-13.
3. See Kilbourn, *Pipeline*; Dexter, "Politics, Pipelines and Parliament."
4. See J. N. McDougall, "The National Energy Board and Multinational Corporations," ch. III, fn. 15.
5. *Debates*, Feb. 25, 1955, p. 1530.
6. Davis, *Canadian Energy Prospects*, pp. 166-7.
7. Ibid., pp. 185-6.
8. It is possible, however, to overestimate the role of foreign ownership as such in such pricing arrangements, since export prices were also a consequence of the American Federal Power Commission's insistence in the case of Westcoast that the Pacific Northwest market must not be allowed to depend exclusively on gas imported from Canada, a consideration that presumably would have reduced the price obtainable for exports by Canadian-owned companies as well. This stance on the part of the FPC adversely affected the Westcoast export price because these sales had to absorb the cost of extending pipelines from American sources into the same market areas: see Davis, pp. 165-6. Some observers have attributed this position to protectionism urged on the FPC by American gas producers and have suggested that it applied generally to all imports of gas from Canada during the 1950s. However, detailed analysis of the Trans-Canada export proposals during the decade denies that

the national origin of the gas was the problem in that instance. Compare Ian McDougall, "The Canadian National Energy Board," pp. 331-2; and H. G. J. Aitken, "The Midwestern Case," p. 139.

9. Royal Commission on Canada's Economic Prospects, *Final Report* (1957), p. 146.
10. Ibid., pp. 125-7.
11. *Debates*, Feb. 11, 1957, pp. 1155-60.
12. See Order-in-Council P.C. 1957, 1386. This order is reproduced as an appendix to Ian McDougall, "The Canadian National Energy Board," at p. 381.
13. Royal Commission on Energy, Transcripts of Briefs and Statements, I-LX, p. 429 (hereafter cited as *Hearings*).
14. Ibid., pp. 2467-9.
15. Ibid., pp. 3129, 3143, and 3137.
16. Ibid., p. 6751.
17. Ibid., pp. 3622 and 3655.
18. Ibid., p. 902 (CPA); pp. 942-4 (Westcoast); p. 4473 (Shell).
19. An interesting reflection of this relationship between imports and prices is found in one of the briefs supporting exports, which lauded the benefits gas exports would bestow on Alberta's coal industry when the price of natural gas in Western Canada eventually rose to a level where coal producers could capture Canadian markets they hadn't enjoyed for years: see ibid., pp. 646 and 674.
20. Ibid., pp. 7164 and 7195.
21. Royal Commission on Energy, *First Report*, p. vii.
22. Ibid., p. 8.
23. Ibid., pp. 7-8.
24. Ibid., p. vii.
25. Ibid., p. 10.
26. See, for example, ibid., pp. 10-11.
27. Ibid., p. 11. This idea, incidentally, was finally to be accepted by the federal government as part of the National Energy Program of 1980.
28. Ibid., pp. 11-12 and p. vii.
29. S.C. 1955, c. 14. Some commentators see the original version of this act and the conditions that obtained regarding natural gas exports from southwestern Ontario at the turn of the century as a precedent for many of the problems later encountered with respect to the export of gas from Western Canada. Compare Ian McDougall, "The Canadian National Energy Board," pp. 334-5; Miller, *Foreign Trade in Gas and Electricity*, pp. 75-76; The Electricity and Fluid Exportation Act. S.C. 1907, c. 16. I have not included this act and the political debate arising from it in my review of policies with respect to the trade and transportation of Canadian

fuels for two reasons: first, the volumes of natural gas involved were very small in relation to Ontario's fuel needs at the time; second, both the government and the opposition were almost exclusively preoccupied with the export, not of natural gas, but of electrical power during the course of two sessions of debate on the original legislation. There are, however, some interesting parallels between this early debate on the merits of exporting power from Ontario and later ones on the export of natural gas from Canada.

30. Quoted by the commission in its *First Report*, p. 13.
31. Compare ibid., pp. ix-xi. For the remainder of this chapter, compare the reports and recommendations of the commission with The National Energy Board Act, S.C. 1959, c. 46, esp. ss. 44 and 83.
32. Ibid., p. xii.
33. Ibid.
34. See Royal Commission on Energy, *Second Report*, pp. 5.15-16.
35. *Hearings*, pp. 4124 and 4126.
36. *Second Report*, p. 5.16. See also *Hearings*, p. 4160.
37. *Hearings*, pp. 4792-4808 (Imperial); 4491-9 (Shell); 5063 (McColl-Frontenac); 5318 (BA); 5656 (California Standard); 7731 (Sun); 7808 (Petrofina).
38. Ibid., p. 4287-94. See also *Second Report*, p. 5.17.
39. On most of these points, see the oil company testimony cited in note 37. See also *Hearings*, pp. 4564 and 5123.
40. Ibid., pp. 5882-5910; 5484-5512.
41. See Davis, *Canadian Energy Prospects*, pp. 126-7; Aitken, *The American Economic Impact*, p. 25.
42. See *Second Report*, ch. 6.
43. Ibid., pp. 6.18-19.
44. Ibid. The *Second Report* was issued about the same time the NEB was established.
45. *Debates*, Aug. 26, 1958, p. 4143.
46. Ibid., p. 4144.
47. Ibid., March 11, 1959, p. 1831.
48. Ibid., Aug. 26, 1958, p. 4023.
49. Ibid., March 11, 1959, p. 1831.
50. Ibid., July 18, 1958, p. 2373.
51. Ibid., p. 2376.
52. *First Report*, p. vii.
53. Ibid., p. 8.
54. Ibid., p. 11.
55. *Debates*, May 18, 1959, p. 3771-4; 3783; May 25, 1959, 3980-98; 4002-5.
56. Ibid., June 1, 1959, p. 4218.
57. Ibid., Feb. 1, 1961, pp. 1641-2.

58. Ibid., Aug. 10, 1960, p. 7897.
59. Ibid., Feb. 1, 1961, pp. 1642-3.

6. Natural Gas and the National Interest: National Energy Board Decisions from 1960–1971

1. National Energy Board Act, S.C. 1959, c. 46, s. 83.
2. For a detailed discussion of the board's calculation of surplus, see Ian McDougall, "The Canadian National Energy Board," pp. 358-65.
3. NEB, *Annual Report for the Year Ended 31 December 1969*, pp. 12-13.
4. Transcript of the hearing of the National Energy Board commencing Nov. 25, 1969, and ending March 20, 1970, into matters reported in the report to the Governor-in-Council of August 1970, p. 1293. Hereafter cited as *Board Hearings*.
5. Ibid., pp. 5845-7.
6. Ibid., p. 1859.
7. Ibid., p. 1860.
8. Ibid., exhibits 68 (Gulf) and 70 (Shell); pp. 5862-3 (Amoco); p. 5882 (Amerada Hess); and exhibit 83 (Mobil).
9. Ibid., p. 5876.
10. Ibid., pp. 5890 (B.C. Hydro); 5941 (Northern and Central); 5956-60 (Gaz Metropolitain); 5961 (Consumers' Natural Gas); 5979-85 (Union Gas).
11. Ibid., pp. 1556-9 (Trans-Canada); 5760 (Westcoast).
12. Ibid., p. 6039.
13. Ibid., p. 5890.
14. Ibid., p. 5863.
15. Ibid., p. 5900.
16. Ibid., pp. 5856; 5858-9; 5862-3; 5882-4; 5898-5900; exhibits 70 and 83. Gulf expressed its support of the CPA position in exhibit 68.
17. Ibid., exhibit 117.
18. Ibid., pp. 5638 and 5991-4.
19. Ibid., exhibit 81.
20. NEB, *Report to the Governor-in-Council*, Aug. 1970, p. 10.18.
21. Ibid., p. 10.19.
22. Ibid., p. 5.31.
23. Such an account is available in J. N. McDougall, "The National Energy Board and Multinational Corporations," pp. 113-35.
24. The decisions of the NEB with respect to these two firms and their export pricing arrangements are reviewed in Ian McDougall, "The Canadian National Energy Board," pp. 347-53.

25. NEB *Report*, Aug. 1970, pp. 10.26; cf. p. 5.35.
26. Ibid., p. 10.28.
27. Ibid., pp. 10.25-30.
28. Ibid., p. 10.31.
29. El Paso at this time owned 17 percent of Westcoast. For an assessment of the implications of intercorporate ownership for the decisions discussed here, see Ian McDougall, "The Canadian National Energy Board," pp. 349-51.
30. NEB *Report to the Governor-in-Council*, March 1967, pp. 8.10-15.
31. NEB, *Report to the Governor-in-Council*, Feb. 1968, p. 9.
32. Ibid., p. 10.
33. Quoted by the board, ibid., p. 4.
34. NEB *Report to the Governor-in-Council*, Aug. 1970, pp. 9.39-40 and 10.60.
35. Ibid., p. 10.62.
36. Ibid., pp. 10.63-4.
37. National Energy Board Act, supra, n. 1, s. 44.
38. NEB *Report to the Governor-in-Council*, July 1961, p. 16.
39. Ibid., p. 36.
40. Ibid., pp. 39-46.
41. Ibid., p. 56.
42. NEB *Report to the Governor-in-Council*, Aug. 1966, p. 6.7.
43. Ibid., p. 6.12.
44. Ibid.
45. *Board Hearings*, pp. 5842-3; 5850.
46. Ibid., p. 5856.
47 Ibid., pp. 5860-61 (IPAC); 5862-6 (Amoco); 5868-71 (Banff Oil); 5875-7 (Canadian Fina); 5882-5 (Amerada Hess); 5898-5900 (Dome); and exhibits 68 (Gulf), 70 (Shell), and 83 (Mobil).
48. Ibid., pp. 5991-6002.
49. Ibid., pp. 6050-51.
50. Ibid., pp. 6561-2.
51. Ibid., p. 5981 (Union); 5969 (Consumers').
52. Ibid., pp. 5771-2 (Trans-Canada); 5832-3 (Westcoast); 5940 (Northern and Central); 6013-17 (Manitoba); 6029 (Ontario); and 6042-3 (Quebec).
53. Ibid., pp. 5887, 5916-28, and 5960.
54. Ibid., pp. 5796 (Trans-Canada); 5964 (Consumers'); 5941-3 (Northern and Central); 6027-8 (Ontario); 6041-2 (Quebec); and 6051 (Saskatchewan Power Corporation).
55. NEB *Report to the Governor-in-Council*, Aug. 1970, pp. 10.43-4.
56. It has been argued that this "export orientation" of the board is evidence of its "capture" by the industry it regulates. However, it is important to bear in mind that the background of most members

of the board has been in the public service rather than the industry, and to note Lucas and Bell's conclusion that, for this and related reasons, "the Board more often views the world through the eyes of the executive branch of government than of the industry." See their *The National Energy Board*, pp. 40-41.

57. Kilbourn, *Pipeline*, p. 182. Kilbourn reviews the Great Lakes affair at length in ch. 11. of his book.
58. Ibid., pp. 182-3. See also *Debates*, Oct. 28, 1966, p. 9223.
59. *Debates*, Oct. 28, 1966, pp. 9224-5.
60. Ibid., p. 9227.
61. Ibid., p. 9610.
62. For example, see *Debates*, Oct. 31, 1966, pp. 9327, 9339-42, 9465.
63. For example, see *Debates*, Oct. 28, 1966, pp. 9228, 9509; Nov. 7, 1966, 9613; Nov. 16, 1966, 9964.
64. The parliamentary reaction (if any) to all of the board's decisions during this period has been related in J. N. McDougall, "The National Energy Board and Multinational Corporations," ch. V.
65. See, for example, James Laxer's account of the emergence of a "continental energy strategy" in *Canada's Energy Crisis*, ch. 8.
66. See *Debates*, Oct. 9, 1970, pp. 22-29.
67. Ibid., Oct. 5, 1970, p. 8755.
68. Ian McDougall, "The Canadian National Energy Board."
69. *Debates*, April 1, 1960, p. 2853.

7. National Self-sufficiency and North American Interdependence: Canadian Energy Policy in the 1970s

1. For a review of the changing circumstances surrounding the Northern Pipeline project and the government's different lines of argument in its defence over time, see Bregha, *Bob Blair's Pipeline*.
2. Address to Symposium, Society of Petroleum Engineers, March 9, 1971, p. 11. Quoted in Askew, *Continentalism Versus Nationalism*, pp. 73-4.
3. Dosman, *The National Interest*, pp. xv and 24. As pointed out above with respect to the Matador Pipeline case, it seems contradictory that Canada should try to improve the market prospects for Canadian oil and gas by contributing to the capacity of the United States to serve the same markets with its own supplies. Once again, however, the argument was that such cooperative efforts on Canada's part would improve the general climate of Canada-U.S. energy relations on which Canadian exports ultimately depended.
4. *Debates*, Dec. 6, 1973, pp. 8478, 8482.
5. See Bregha, esp. chs. 5-7.
6. For a comprehensive account of the politics surrounding the vic-

tory of the Foothills proposal over that of CAGPL, see ibid.

7. It is curious that numerous critics of the NEB and the government have argued that the NEB overestimated Canada's exportable surplus of gas, so as to promote increased exports in 1970, but relatively few who have considered that the NEB and the government may have underestimated reserves in Alberta after 1974 in order to promote the construction of the Mackenzie Pipeline as a necessary means of avoiding apparently looming gas shortages. For evidence of this possibility, see John Helliwell, "Arctic Pipelines in the Context of Canadian Energy Requirements," esp. p. 350. See also Bregha, pp. 58-60.

8. For a notable exception, see Helliwell, "Impact of a Mackenzie Pipeline on the National Economy."

9. The extraordinary character of the Berger Inquiry's procedures, report, and recommendations are reviewed in Bregha, pp. 115-23. Strictly speaking, Berger was charged only with the responsibility to inquire into the "terms and conditions" under which a northern pipeline would be built, but he reported on these matters in such a way that the original Mackenzie Valley proposal was politically doomed.

10. *Debates*, Feb. 20, 1978, p. 3017.

11. Ibid., Feb. 13, 1978, p. 2788.

12. Ibid., p. 2812.

13. See Douglas's closing speech on the day the vote on C-25 was taken: ibid., April 4, 1978, p. 4142.

14. Ibid., Feb. 13, 1978, p. 2794.

15. Ibid., p. 2810.

16. Ibid., March 22, 1978, pp. 4037-40. The division is recorded on April 4, 1978, p. 4080-1.

17. Ibid., April 4, 1978, pp. 4149-50.

18. Ibid., Feb. 20, 1978, pp. 3020-1.

19. Ibid., March 22, 1978, p. 4037.

20. Ibid., April 4, 1978, p. 4140.

21. Government concern to supply Montreal with domestic oil had been activated in early 1973, before the Middle East War and the Arab oil embargo. However, these events clearly intensified this concern and gave greater impetus to the Interprovincial extension. See Lucas and Bell, *The National Energy Board*, pp. 79-80.

22. E.g., Stanfield, *Debates*, Jan. 7, 1974, p. 9109; Saltsman, Oct. 25, 1973, p. 7241.

23. See Lucas and Bell, p. 92. These concerns of the industry were raised in opposition to the pipeline by some Conservative members, whose party did not always speak with a single voice on the issue, even in consecutive speeches. See, for example, Gordon

Ritchie, *Debates*, Nov. 4, 1974, p. 1036, Cf. Sinclair Stevens, p. 1033.

24. See McRae, *Debates*, Oct. 25, 1973, p. 7238; Saltsman, p. 7241.

25. Ibid., Oct. 25, 1973, p. 7223.

26. See, for example, Hamilton, *Debates*, Nov. 13, 1974, pp. 293-4; Stanfield, Nov. 14, 1978, pp. 1344-5.

27. Macdonald, ibid., October 31, 1974, p. 914.

28. Ibid.

29. Ibid., Nov. 14, 1974, p. 1345.

30. Ibid.

31. Ibid., Dec. 5, 1974, p. 1977, cf. Stanfield, ibid., Nov. 14, 1974, p. 1344.

32. Ibid., Nov. 7, 1974, p. 1166. cf. Andre, ibid., Nov. 4, 1974, p. 1032.

33. Ibid., Oct. 31, 1974, p. 922.

34. Ibid., April 14, 1975, pp. 4775-6.

35. Ibid., Dec. 5, 1974, p. 1979.

36. Ibid., p. 1989.

37. Ibid., Nov. 1, 1974, p. 983. One academic analysis estimates the "average avoidable annual cost to consumers" of the National Oil Policy from 1963 to 1970 at $75 million, which would tend to support the lower of the two figures given by Saltsman. See Anderson, "Price Formation in the Canadian Crude Oil Sector," p. 11.

38. Strictly speaking, the amendment was to condition 12 of schedule 3 of the act, not to the act itself.

39. An excellent detailed analysis of the pre-build decision is available in Bregha, ch. 11.

40. "Quebec gas pipeline may be delayed," *Oilweek*, August 11, 1980. This article, ironically in the issue containing a feature report on the pre-build decision, ends on a note of pessimism concerning the viability of the eastern extension. The NEB postponed consideration on the Maritimes line pending the completion of further environmental studies and a fuller assessment of the fuel supply potential off the East Coast.

41. The argument presented in the remainder of this section does not proceed from the assumption that the construction of the southern sections of the northern pipeline will not, contrary to the assurances given at the time by the Canadian government, contribute to the eventual completion of the entire project; rather, it is based on the observation that the government abandoned its previous insistence that construction on any part of the Alaska pipeline must not commence until financing had been obtained for the project as a whole. The decision of July 1980 to approve the construction of the southern section entailed an amendment to the

Northern Pipeline Act that, in effect, eliminated the restriction that no construction on the Alaska Highway pipeline commence until financing of the entire line had been obtained, and substituted for it the requirement that there be evidence that the financing of the southern section had been obtained and that the financing of the remainder of the line *could* be obtained. I am therefore arguing simply that there was an element of doubt (and risk) involved in the difference between a pipeline to Alaska for which the financing "has been obtained" and one for which the financing "can be obtained." Compare condition 12 of schedule 3 of the Northern Pipeline Act, S.C. 1977–78, c. 20, and appendix A of the National Energy Board, *Finding*, July 1980. For a succinct summary of the risks involved, see Jeff Carruthers, " 'Prebuild' line: siphon for U.S.," *Globe and Mail*, August 26, 1980. See also Bregha, p. 220.

42. The position of the NDP and other parties concerning the pre-build project can be reviewed in *Debates*, July 31, 1980, pp. 3150-64 and 3172-91. See also Bregha, p. 226.

43. See the communications to the Canadian government on this matter from U.S. President Jimmy Carter, Congress, and the American sponsors of the pipeline, reproduced as appendices F, E, and D respectively, in the NEB's *Finding*, July 1980. See also Bregha, pp. 221-5.

44. Prime Minister Trudeau described the decision as "a judgment call" in Parliament on July 14, 1980.

45. See, respectively, Energy, Mines and Resources Canada, *The National Energy Program, 1980*, pp. 48-52; 42-8; 32-8; and 108-12. Hereafter cited as NEP.

46. NEP, p. 2. Emphasis in original. See also pp. 7-10.

47. Ibid., pp. 56, 58, 81-2.

48. As one telling indication of support, several leading spokesmen for the Committee for an Independent Canada gave the government "full marks" for its new program a short time before the CIC formally disbanded.

49. NEP, pp. 44-5.

50. See NEP, pp. 41-2.

8. The Politics of Energy in Canada: Some Conclusions

1. Simeon, "Studying Public Policy."

2. Ibid., p. 580.

3. Ibid., pp. 566-78.

4. Ibid., p. 565.

5. Ibid., pp. 568-9.

6. Petro-Canada, *1980 Annual Report*, pp. 20-21.

7. Simeon, p. 573.
8. Ibid., pp. 575-7.
9. For a discussion of some of these parallels, see Kellas, "Oil, Federalism and Devolution."
10. Thomson, "Some Economic Aspects of the Canadian Coal Problem," p. 403-4.

Bibliography

Books

Adelman, M. A. *The World Petroleum Market*. Baltimore: Johns Hopkins University Press, 1972.

Adelman, M. A. *The Supply and Price of Natural Gas*. Oxford: Basil Blackwell, 1962.

Adelman, M. A., P. C. Bradley, and C. A. Norman. *Alaskan Oil: Costs and Supply*. New York: Praeger Publishers, 1971.

Aitken, H. G. J. *American Capital and Canadian Resources*. Cambridge: Harvard University Press, 1961.

Aitken, H. G. J., ed. *The American Economic Impact on Canada*. Durham: Duke University Press, 1959.

Askew, J. Coulter. *Continentalism Versus Nationalism: The Party Politics of Oil and Gas Pipelines in Canada, 1949-1976*. M.A. thesis, University of Western Ontario, 1977.

Axline, A., J. E. Hyndman, P. V. Lyon, and M. A. Molot, eds. *Continental Community? Independence and Integration in North America*. Toronto: McClelland and Stewart, 1974.

Bercuson, David, ed. *Alberta's Coal Industry, 1919*. Calgary: Historical Society of Alberta, 1978.

Bregha, Francois. *Bob Blair's Pipeline: The Business and Politics of Northern Energy Development Projects*. Toronto: James Lorimer and Co., 1979.

Brown, Richard. *The Coal Fields and Coal Trade of the Island of Cape Breton*. London: Sampson Low, Morston, Low and Searle, 1871.

Burley, K. H. *The Development of Canada's Staples, 1867-1939: A Documentary Collection*. Toronto: McClelland and Stewart, 1970.

Burton, Thomas. *Natural Resource Policy in Canada*. Toronto: McClelland and Stewart, 1972.

Cicchetti, C. J. *Alaskan Oil: Alternative Routes and Markets*. Baltimore: Johns Hopkins University Press, 1972.

Creighton, Donald. *The Road to Confederation: The Emergence of Canada, 1863-1867*. Cambridge: Houghton Mifflin, 1965.

Cuff, R. D. and J. L. Granatstein. *Canadian American Relations in Wartime*. Toronto: Hakkert, 1975.

Davis, E. M. *Canada's Oil Industry*. Toronto: McGraw-Hill, 1969.

Davis, John. *Natural Gas and Canadian-American Relations*. Washington: Canadian-American Committee, 1959.

Davis, John. *Oil and Canada-United States Relations*. Washington: Canadian-American Committee, 1959.

Dosman, E. *The National Interest: The Politics of Nothern Development 1968-1975*. Toronto: McClelland and Stewart, 1975.

Easterbrook, W. T., and M. H. Watkins, eds. *Approaches to Canadian Economic History*. Toronto: McClelland and Stewart, 1967.

Engler, Robert. *The Politics of Oil*. Chicago: University of Chicago Press, 1961.

Erickson, E., and L. Waverman. *The Energy Question: An International Failure of Policy*. Vol. 2. Toronto: University of Toronto Press, 1974.

Evans, Douglas. *Western Energy Policy: The Case for Competition*. London: Macmillan, 1978.

Forsey, Eugene. *National Problems of Canada: Economic and Social Aspects of the Nova Scotia Coal Industry*. Toronto: Macmillan, 1929.

Grant, G. *Technology and Empire*. Toronto: House of Anansi, 1970.

Gray, Earle. *The Impact of Oil: The Development of Canadian Oil Resources*. Toronto: Ryerson Press, 1969.

Gray, Earle. *The Great Canadian Oil Patch*. Toronto: Maclean-Hunter, 1970.

Hanson, E. J. *Dynamic Decade: The Evolution and Effects of the Oil Industry in Alberta*. Toronto: McClelland and Stewart, 1958.

Hilborn, James, ed. *Dusters and Gushers: The Canadian Oil and Gas Industry*. Toronto: Pit Publishing Co., 1968.

Hopkins, J. Castel. *The Canadian Annual Review of Public Affairs, 1922*. Toronto: The Canadian Review Company, 1923.

Innis, H. A. *Select Documents in Canadian Economic History, 1497-1783*. Toronto: University of Toronto Press, 1929.

Innis, Mary Q. *An Economic History of Canada*. Toronto: Ryerson Press, 1935.

Kilbourn, W. *Pipeline: Trans-Canada and the Great Debate, A History of Business and Politics*. Toronto: Clarke Irwin Co., 1970.

Krislov, S., and L. D. Musolf, eds. *The Politics of Regulation: A Reader*. Boston: Houghton Mifflin Co., 1964.

Laxer, James. *The Energy Poker Game: The Politics of the Continental Resources Deal*. Toronto: New Press, 1970.

Laxer, James. *Canada's Energy Crisis*. Toronto: James Lorimer, 1975.

Laxer, J., and A. Martin, eds. *The Big Tough Expensive Job: Imperial Oil and the Canadian Economy*. Toronto: Press Porcepic, 1976.

Levitt, K. *Silent Surrender: The Multinational Corporation in Canada*. Toronto: Macmillan, 1969.

Lewis, D. E., and A. R. Thompson. *Canadian Oil and Gas*. Toronto: Butterworths, 1955.

Lucan, A. R., and T. Bell. *The National Energy Board: Policy, Procedure, and Practice*. Ottawa: Law Reform Commission of Canada, 1977.

McDougall, J. N. *The National Energy Board and Multinational Corporations: The Politics of Pipelines and Natural Gas Exports, 1960-1971*. Ph.D. thesis, University of Alberta, 1975.

Miller, John T. *Foreign Trade in Gas and Electricity in North America*. New York: Praeger, 1970.

Muise, D. A. *Elections and Constituencies: Federal Politics in Nova Scotia, 1867-1878*. Ph.D. thesis, University of Western Ontario, 1971.

Naylor, T. *The History of Canadian Business*. Vols. I and II. Toronto: James Lorimer & Co., 1975.

Nelles, H. V. *The Politics of Development: Forests, Mines and Hydro-Electric Power in Ontario, 1849-1941*. Toronto: Macmillan, 1974.

Newton, D. H., and N. V. Brechner. *The Oil Security System*. Lexington, Mass.: D. C. Heath and Company, 1975.

Nordeg, Martin. *The Fuel Problems of Canada*. Toronto: Macmillan, 1930.

Pearse, Peter, ed. *The Mackenzie Valley Pipeline*. Toronto: McClelland and Stewart, 1974.

Penrose, E. T. *The Large International Firm in Developing Countries: The International Petroleum Industry*. London: George Allen and Unwin.

Plotnick, A. R. *Petroleum: Canadian Markets and United States Foreign Trade Policy*. Seattle: University of Washington Press, 1964.

Pratt, L. R. *The Tar Sands: Syncrude and the Politics of Oil*. Edmonton: Hurtig Publishers, 1976.

Richards, John and L. R. Pratt. *Prairie Capitalism. Power and Influence in the New West*. Toronto: McClelland and Stewart, 1979.

Rogers, G. W., ed. *Change in Alaska: People, Petroleum and Politics*. Fairbanks: University of Alaska Press, 1970.

Ross, Victor. *Petroleum in Canada*. Toronto: Southam Press, 1917.

Sykes, P. *Sellout: The Giveaway of Canada's Energy Resources*. Edmonton: Hurtig Publishers, 1973.

Szyliowicz, J. S., and R. E. O'Neill, eds. *The Energy Crisis and United States Foreign Policy*. New York: Praeger, 1975.

Tanzer, M. *The Political Economy of International Oil and the Underdeveloped Countries*. Boston: Beacon Press, 1969.

Waite, P. B. *The Life and Times of Confederation, 1864-1867: Politics, Newspapers and the Union of British North America*. Toronto: University of Toronto Press, 1962.

Waverman, L. *Natural Gas and National Policy*. Toronto: University of Toronto Press. 1973.

Winn, C., and J. McMenemy, eds. *Political Parties in Canada*. Toronto: McGraw-Hill Ryerson, 1976.

Articles

Acheson, T. W. "The National Policy and the Industrialization of the Maritimes, 1880-1910." *Acadiensis* 1 (1972): 3-28.

Adelman, M. A. "The Alaskan North Slope Discoveries and World Petroleum Supplies and Costs," in G. W. Rogers, ed., *Change in Alaska: People, Petroleum and Politics*. Fairbanks: University of Alaska Press, 1970.

Aitken, H. G. J. "The Changing Structure of the Canadian Economy," in Aitken (ed.) *American Economic Impact on Canada*. Durham: Duke University Press, 1959.

Aitken, H. G. J. "The Midwestern Case: Canadian Gas and the Federal Power Commission." *Canadian Journal of Economics and Political Science* 25 (1959): 129-43.

Anderson, F. J. "Price Formation in the Canadian Crude Oil Sector." *Canadian Public Policy* 2 (1976): 1-16.

Bradley, P. G. "Canadian Energy Policy: Some Economic Questions." *B.C. Studies* 13 (1972): 110-200.

Berry, G. R. "The Oil Lobby and the Energy Crisis." *Canadian Public Administration* 17 (1974): 600-636.

Carter, R. C. "The National Energy Board of Canada and the American Administrative Procedure Act: A Comparative Study." *Saskatchewan Law Review* 34 (1969): 104-42.

Dales, J. H. "Fuel, Power and Industrial Development in Central Canada." *American Economic Review, Papers and Proceedings* 43 (1953): 181-98.

Debanne, J. G., "Oil and Canadian Policy," in E. Erickson and L. Waverman, eds., *The Energy Question: An International Failure of Policy*, vol. 2. Toronto: University of Toronto Press, 1974.

Dexter, Grant. "Politics, Pipelines and Parliament." *Queens Quarterly* 63 (1956): 323-33.

Fisher, B. "The Role of the National Energy Board in Controlling the Export of Natural Gas from Canada." *Osgoode Hall Law Journal* 9 (1971): 553-601.

Foster, Ben. "The Coming of the National Policy: Business, Government and the Tariff, 1876-1879." *Journal of Canadian Studies* 14 (1979): 39-49.

Fowke, V. C. "The National Policy: Old and New." *Canadian Journal of Economics and Political Science* 38 (1952): 271-86.

Gainer, W. D. and T. L. Pourie. "Public Revenue from Canadian Crude Petroleum Production." *Canadian Public Policy* 1 (1975): 1-12.

Gray, F. W. "Fifty Years of the Dominion Coal Co." *Dalhousie Review* 22 (1943): 460-69.

Greenwood, Ted. "Canadian-American Trade in Energy Resources." *International Organization* 28 (1974): 689-711.

Hamilton, Richard E. "A Marketing Board to Regulate Exports of Natural Gas." *Canadian Public Administration* 16 (1973): 83-96.

Hamilton, R. G. "Natural Gas and Canadian Policy," in F. Erickson, and L. Waverman, *The Energy Question: An International Failure of Policy.* Toronto: University of Toronto Press, 1974.

Hanson, E. J. "Natural Gas in Canadian-American Relations." *International Journal* 12 (1957): 186-99.

Helliwell, John. "Impact of a Mckenzie Pipeline on the National Economy," in Peter H. Pearse, ed., *The Mackenzie Pipeline: Arctic Gas and Canadian Energy Policy.* Toronto: McClelland and Stewart, 1974.

Helliwell, John. "Arctic Pipelines in the Context of Canadian Energy Requirements." *Canadian Public Policy* 3 (1977): 344-54.

Innis, H. A. "The Canadian Mining Industry," in Mary Q. Innis, ed., *Essays in Canadian Economic History.* Toronto: University of Toronto Press, 1956.

Innis, H. A. "Transportation as a Factor in Canadian Economic History," in Mary Q. Innis, ed., *Essays in Canadian Economic History.* Toronto: University of Toronto Press, 1956.

Kellas, James G. "Oil, Federalism and Devolution: A Canadian-British Comparison." *The Round Table* 259 (1975): 273-80.

Kierans, E. "The Day the Cabinet Was Misled." *Canadian Forum*, March 1974.

Lindberg, Leon. "Energy Policy and the Politics of Economic Development." *Comparative Political Studies* 10 (1977): 355-82.

MacAvoy, P. W. "The Regulation-Induced Shortage of Natural Gas." *Journal of Law and Economics* 14 (1971): 167-99.

McDougall, Ian. "The Canadian National Energy Board: Economic 'Jurisprudence' in the National Interest or Symbolic Reassurance?" *Alberta Law Review* 22 (1973): 327-82.

McDougall, Ian. "Canada's Oil and Gas: An Eleventh Hour Option that Must Not Be Ignored." *Canadian Public Policy* (1975) 1: 47-57.

McDougall, J. N. "Oil and Gas in Canadian Energy Policy," in B. Doern and S. Wilson, *Issues in Canadian Public Policy.* Toronto: Macmillan, 1974.

McDougall, J. N. "Regulations vs. Politics: The National Energy Board and the Mackenzie Valley Pipeline," in A. Axline et al., eds., *Continental Community? Independence and Integration in North America.* Toronto: McClelland and Stewart, 1974.

Page, D. M. "An Energy Crisis in Reverse: Canada as a Net Oil Importer." *International Perspectives* (March/April 1974): 18-21.

Pratt, Larry. "Accountability: The State and Oil." *Canadian Forum*, August 1976.

Roseman, F. and R. W. Wilkinson. "Who Benefits? The Alberta Energy Price Increases." *Canadian Forum,* June-July 1973.

Scott, A. "Policy for Crude Oil." *Canadian Journal of Economics and Political Science* 27 (1961): 267-76.

Simeon, Richard. "Studying Public Policy." *Canadian Journal of Political Science* 19 (1976): 548-80.

Thomson, L. R. "Some Economic Aspects of the Canadian Coal Problem," in *Proceedings of the Symposium on Fuel and Coal.* Montreal: McGill University Press, 1931.

Thorburn, H. G. "Parliament and Policy-Making: The Case of the Trans-Canada Gas Pipeline." *Canadian Journal of Economics and Political Science* 23 (1957): 516-31.

Waverman, L. "Energy in Canada: A Question of Rents," in L. H. Officer and L. B. Smith, eds., *Issues in Canadian Economics.* Toronto: McGraw-Hill Ryerson, 1974.

Waverman, L. "National Policy and Natural Gas: The Costs of a Border." *Canadian Journal of Economics* 5 (1972): 331-49.

Waverman, L. "The Two Price System in Energy: Subsidies Forgotten." *Canadian Public Policy* 1 (1975): 76-86.

Waverman, L. "The Reluctant Bride: Canadian-American Energy Relations," in E. Erickson, and L. Waverman, eds., *The Energy Question: An International Failure of Policy,* vol. 2. Toronto: University of Toronto Press, 1974.

Yanchula, Joseph. "The Politics of Petroleum: An Inside Look at Alberta Oil." *Canadian Forum,* Nov.-Dec. 1974.

Government Documents

Alberta

Oil and Gas Conservation Board. *Gas Export 1950-1960.* A summary of the proceedings before the Oil and Gas Conservation Board pursuant to the provisions of the Gas Resources Preservation Act. Compiled by D. P. Goodall. Calgary: Queen's Printer, 1961.

Provincial Archives of Alberta. *Premiers' Papers.*

Royal Commission to Inquire into Matters Connected with Petroleum and Petroleum Products (the McGillivray Commission). *Report.* Edmonton: King's Printer, 1946.

Royal Commission Respecting the Coal Industry (the Barlow Commission). *Report* (1935). Edmonton: King's Printer, 1936.

Canada

Department of Defence Production, Economics and Statistics Branch. *The Natural Gas Industry.* Ottawa: 1951.

Department of Energy, Mines and Resources. *An Energy Strategy for Canada: Policies for Self-Reliance.* Ottawa: Queen's Printer, 1976.

Department of Energy, Mines and Resources. *The National Energy Program, 1980.* Ottawa: Supply and Services Canada, 1980.

Dominion Fuel Board. *Interim Report of the Dominion Fuel Board.* Ottawa: 1923.

Dominion Fuel Board. *Second Progress Report, 1923-1928*. Ottawa: 1928.

Final Report of the Fuel Controller, Canada, 1919. Ottawa: King's Printer, 1919.

National Energy Board. *Annual Report for the Year Ended 31 December 1969*. Ottawa: Information Canada, 1970.

National Energy Board. *Reports to the Governor-in-Council in the Matter of Applications under the National Energy Board Act*. Ottawa: July 1961, August 1966 and August 1970.

Parliament of Canada. *Evidence taken respecting the future fuel supply of Canada*. Ottawa: 1921.

Parliament of Canada. *Final Report of Special Committee on the Future Fuel Supply of Canada*. Ottawa: 1921.

Parliament of Canada. *Final Report of a Special Committee on the Fuel Supply of Canada*. Ottawa: 1923.

Parliament of Canada. *Third and Final Report of the Select Committee on Mines and Minerals*. Ottawa: 1923.

Public Archives of Canada. *C. D. Howe Papers*.

Public Archives of Canada. *W. L. M. King Papers*.

Royal Commission on the Coal Industry in Canada (the Carroll Commission). *Report*. Ottawa: King's Printer, 1947.

Royal Commission on Energy (the Borden Commission). *First Report*. Ottawa: Queen's Printer, 1958.

Royal Commission on Energy (the Borden Commission). *Second Report*. Ottawa: Queen's Printer, 1959.

Index